A BURST OF
CONSCIOUS
LIGHT

"In this intriguing book, Andrew Silverman shows that consciousness is fundamental and not merely the product of a brain. Many eminent scientists have been driven to this conclusion by evidence from quantum physics. This is also supported by evidence from near-death experiences (NDEs) in which clear conscious perceptions continue while the brain is not functioning. During NDEs many people experience a very bright light that emanates pure and unconditional love, undivided knowledge, and universal interconnectedness. This groundbreaking book presents cogent scientific evidence that this same light is the fundamental basis for existence itself and that we all contain a spark of this light. A very original book that is based on up-to-date interpretations of modern science. Highly recommended."

PIM VAN LOMMEL, CARDIOLOGIST AND
AUTHOR OF *CONSCIOUSNESS BEYOND LIFE*

"The modern scientific view of the world is shifting rapidly due to the converging evidence of a mental universe from quantum physics and neuroscience. Dr. Silverman's brilliant opus represents an important milestone in catalyzing this world awakening, which ultimately will weave science and spirituality together again for the betterment of all humankind!"

EBEN ALEXANDER, M.D., NEUROSURGEON AND
AUTHOR OF *LIVING IN A MINDFUL UNIVERSE,*
PROOF OF HEAVEN, AND *THE MAP OF HEAVEN*

A BURST OF
CONSCIOUS
LIGHT

❖

NEAR-DEATH EXPERIENCES,
THE SHROUD OF TURIN,
AND THE
LIMITLESS POTENTIAL OF HUMANITY

DR. ANDREW SILVERMAN

Park Street Press
Rochester, Vermont

Park Street Press
One Park Street
Rochester, Vermont 05767
www.ParkStPress.com

Text stock is SFI certified

Park Street Press is a division of Inner Traditions International

Cataloging-in-Publication Data for this title is available from the Library of Congress

ISBN 978-1-62055-963-5 (print)
ISBN 978-1-62055-964-2 (ebook)

Printed and bound in the United States by Lake Book Manufacturing, Inc.
The text stock is SFI certified. The Sustainable Forestry Initiative® program
promotes sustainable forest management.

10 9 8 7 6 5 4 3 2 1

Text design and layout by Priscilla Baker
This book was typeset in Garamond Premier Pro, with Edwardian Script, Futura,
Nocturne Serif, and Gill Sans used as display typefaces

To send correspondence to the author of this book, mail a first-class letter to the
author c/o Inner Traditions • Bear & Company, One Park Street, Rochester, VT
05767, and we will forward the communication.

Contents

Foreword

Daniel Langsman

Throughout history, people have described themselves and the world around them by using concepts drawn from their everyday experience. Around the time of the Industrial Revolution, the universe and the human body were believed to be like mechanical clocks. More recently, since the information-technology revolution, many people assume that our minds are merely the product of information being processed by the brain, just like a computer. There is even speculation that our minds could be uploaded to machines so that we might transcend mortality.

Many in the scientific community are now raising concerns about this trend, and research institutes have been set up in the premier seats of learning in the world, including the Universities of Oxford and Cambridge, where crucial questions about humanity and its prospects are being discussed by some of the world's leading intellects. These leading minds, including Professor Stephen Hawking, have suggested that the biggest threat to humanity may not be asteroids, nuclear war, or climate change, but rather the unanticipated effects of technology when combined with artificial intelligence. With evidence centered in reason and empirical science, *A Burst of Conscious Light* demonstrates how and why the human mind cannot be reduced to data or software. If we fail to consider the warnings such as those from the Oxford and Cambridge teams, then the continued existence of the human race will indeed be in peril.

Scientific research has demonstrated that people have continued to be conscious during cardiac arrest—when their brain waves were flat and the people were clinically dead. Many scientists and medical doctors are now being forced to the conclusion that our minds may not be the product of our brains and that therefore death may be a comma rather than a full stop. This evidence suggests that the continuation of consciousness beyond death is part of a natural process that has nothing to do with technology. *A Burst of Conscious Light* suggests how this continuation might work. Dr. Andrew Silverman shows that the clues to this have been around for nearly a hundred years and were glimpsed by the founders of quantum mechanics.

There is also more palpable evidence of such continuation by way of a photographic negative image on a cloth that once wrapped a dead body. The cloth can be traced back to the first century and has in more recent times been known as the Shroud of Turin. The subject was previously obscured by a carbon-dating test carried out in the late 1980s. However, many people seem unaware that a senior scientist from Los Alamos National Laboratory in the United States published a scientific paper demonstrating that the corner of the cloth that was assessed in the carbon dating consisted mostly of much newer material that had been introduced in a sixteenth-century repair. This renders the carbon dating result meaningless.

This book details the nature of the image on the Shroud, citing the peer-reviewed published evidence of scientists based at NASA and Los Alamos. It is evident from the work of forensic pathologists that the cloth once wrapped a recently deceased corpse. Research by scientists at the Italian National Agency for New Technologies, Energy, and Sustainable Economic Development (ENEA) suggests that the image was formed on the cloth by a burst of light that emanated from the body that it wrapped. These scientists have estimated that the formation of the image in this way would have required around thirty-four-thousand-billion watts. This is the equivalent of a nuclear blast between two layers of linen cloth. While religious institutions have made this object a focus of worship, Silverman is at great pains to point out that he instead sees it as a focus for scientific enquiry.

An understanding of light is crucial to an understanding of modern physics. Empirical studies show that light also features prominently in near-death experiences. Scientific research implicates light in the formation of the body image on the Shroud. So far, attempts to understand each of these three fields in isolation have reached dead ends. The role of the consciousness of the observer in quantum mechanics is a minefield for scientists. Again, the nature of consciousness in the presence of a lifeless brain is a conundrum for neuroscientists. The scientific explanation for a momentary burst of light from a corpse to produce the image on the Shroud has remained elusive. Until now.

This book provides a unifying way to demonstrate how these three fields of enquiry complete each other within a rational scientific framework. It is written in an engaging style that is accessible to people without a background in science. The author's secular, holistic approach enables him to see connections that others might have missed if they have sought answers in orthodoxy or dogma rather than reason.

Developments in science and technology, although they have brought many benefits for humanity, have also tended to narrow our thinking. They have made us compartmentalized in our outlook. As more information has been revealed by each of the sciences, people have tended to study specialized areas more but without a sense of the overall picture. More detailed knowledge has emerged, but there is less emphasis on how it all might fit together to provide a unified understanding of ourselves and the world we live in.

With over thirty years of experience studying each of these "compartments," the author shares his insights into a view of the world and of humanity without boundaries. His role as a family doctor allows him to share insights that would not have been so easy to gain had he been a different type of scientist. Health care in its true sense involves overseeing the welfare of an individual—psychologically as well as physically. To do this properly depends on seeing the patient as a person with hopes, dreams, and aspirations rather than as just a pile of chemicals in human form. In this way, rather than be an end in itself, clinical science becomes a tool to help the person. This sense of humanity pervades the book. Through reason and a unified approach to the evidence, the book explores how we exist as conscious beings and what the full potential of that consciousness might be.

DANIEL LANGSMAN is a creative artist based in England. He has always been fascinated by the deeper questions of existence and has conducted independent research to help shape his opinions on these subjects. He has collaborated with Dr. Andrew Silverman on many projects over the years and has also attended numerous conferences around the world organized by the Scientific and Medical Network and the Society for Scientific Exploration.

Acknowledgments

I would like to thank the following people without whom this book would not have been possible:

Danielle for her help in arranging opportunities to discuss the ideas in this book with the public, not least through securing an invitation to be interviewed on *Coast to Coast AM*.

Danny for his suggestion that I write this book and for being my sounding board and mentor in many ways as well as for creating some excellent illustrations that are included in this book.

My father (David) for his invaluable support and advice.

Nathan for sharing his profound insights, which are not confined to single malt whiskies but also relate to the human condition in general.

Nigel for being the man referred to in chapter 2 who helped me see that the human "mind's eye" must logically be bigger than the physical universe.

Darren for keeping me grounded, Helen for restoring my faith in humanity, and Julian for pointing out that everyone has their own variety of nonsense, which means that sense and truth do not belong exclusively to anyone.

Lauren for inspiring me with her kindness and empathy.

The marvelous team at Inner Traditions for everything.

Barrie Schwortz, an original STuRP member and its official photographer and the founder of shroud.com, kindly gave permission to include his photos of the STuRP team's groundbreaking research.

Dr. Paolo Di Lazzaro, Joe Marino, and Professor Freeman Dyson for their kind help in clarifying some of their original source material which I have cited.

Last but by no means least, I would like to thank my wife, Lily, for her unwavering love and support.

Author's Note on the Structure of This Book

I hope you will forgive me for using a slightly unconventional paragraph style in this book. When lecturing or giving presentations, I find that I naturally vary the pauses between sentences. I have often used a space before a new line but without a new indent in this book to signify a pause. This should not require any effort on your part as the pause should happen automatically as your eyes track to the next line.

There is a glossary at the end of the book that I have used as a resource to clarify some of the terms I have used in the book.

Before you begin, I'd also like to mention that this book uses sequentially numbered references. This efficient system allows a single note number to be used for the same resource throughout the entire book. I mention this because if you are unfamiliar with the system, it may seem strange if a lower number appears after a higher one on occasion. This just means that there has been an earlier reference to the same item.

A bibliography has also been provided to assist readers who may try to locate sources used in this book. Sequenced alphabetically by author surname, it should alleviate the need to search through the references list.

The Limitless Potential of Humanity

Imagine, if you will, a hypothetical future in which it is proposed that human beings are to be gradually phased out and replaced by machines with "artificial intelligence." These might be devices that in some way attempt to replicate the workings of our brains such that each of us would be replaced by an artificial derivative. These objects, though, would be considered by many people to replicate our memories and personalities and therefore, for all intents and purposes, to be the same as us and even to "be" us.

In 2008, the Future of Humanity Institute at the University of Oxford released a report titled "Whole Brain Emulation: A Roadmap." The authors list a dozen or so "benefits" of whole-brain emulation. They state that:

> If emulation of particular brains is possible and affordable, and if concerns about individual identity can be met, such emulation would enable back-up copies and "digital immortality." . . . [Brain emulation, the authors suggest], may represent a radical new form of human enhancement.[1]

However, before everyone gets carried away with enthusiasm for the idea of artificial immortality, maybe we should consider whether this whole notion of "mind uploading" might be based on a misunderstanding of what consciousness is.

The hypothetical situation that one day people might seek to replace all human beings with "artificially intelligent" machines is one that should lead us to consider whether we might be missing something about us that makes us different from machines. One of the aims of this book is to consider that very question and also to invite you to join me in considering how possible answers might relate to the nature of our very existence as sentient beings. While considering such questions we might find that we have extended our knowledge and understanding of the world to include a clearer understanding of consciousness.

Many people assume that advances in science have relegated the mind to the role of merely an accessory to the brain. On the contrary, many eminent scientists, including many of the founders of the most successful scientific theory ever discovered (quantum theory), didn't see it that way. Erwin Schrödinger, for example, expressed this very clearly:

> Dear reader, recall the bright, joyful eyes with which your child beams upon you when you bring him a new toy, and then let the physicist tell you that in reality nothing emerges from these eyes; in reality their only objectively detectable function is, continually to be hit by and to receive light quanta. In reality, a strange reality! Something seems to be missing in it.[2]

In this book, I will draw your attention to the evidence that shows that consciousness could never be merely a product of the pile of gelatinous substance that we call the brain. The relationship between mind and matter is a fascinating puzzle that I will address in detail. I will show how and where there might be clues to how you can solve it.

Some of the clues are contained in your everyday experience. Not just in the content of your experience but also in the fact that you have a mind that *can* experience. After all, a sound-recording device detects vibration in a way not too dissimilar from how your eardrum detects vibration. Just as you have a brain, modern recording devices have electronic circuitry and software for processing the information in sound. Does a recording device consciously experience the sound? If not, then why do we assume the brain circuitry can produce consciousness?

The view that consciousness may be fundamental is one that is shared by some contemporary eminent scientists notably including physicist Andrei Linde of Stanford University.[3]

What is consciousness, and what does it mean to be aware? Why is there a "you" or an "I" that is able to experience existence? If our minds were made by our brains, which were merely information processors that experienced reality only via the senses and acted according to processes determined solely by physical laws of nature, then why should awareness exist? Just from an evolutionary point of view if nothing else, why should consciousness have conferred any survival advantage if there was no such thing as freedom of will and hence we were merely passive bystanders in existence? It might just as well not be there from this point of view.

People have the sense that they are able to choose freely between available options in any given circumstance. Our notions of personal responsibility assume that we "could have done otherwise" in the sense that we could have made choices other than those we did make, whereas a machine or a computer will have an output determined by the input and by its programming. If there is such a thing as freedom of will, then clearly this would have profound implications for our view of the world and how we relate to it. If freedom of will is "natural," then so is "mind over matter." Surely, if our minds can determine our actions through the exercise of choice, it would follow that we are naturally exerting the power of mind over matter numerous times in our daily lives. If mind over matter is natural, then we no longer need concepts like "supernatural" or "miracle."

Such extraordinary events, if they occur, may not be occasions when the laws of nature are suspended. They may instead provide us with a glimpse into more fundamental and unified laws of nature that apply always but that we usually don't notice because we "cannot see the forest for the trees."

This could give us a clue to help us to understand many phenomena associated with human beings that are as yet unexplained by science. I will present evidence in this book regarding three of the most intriguing scientific mysteries of all time.

The first is the nature and origin of the image on a cloth that is generally known by the name of the Shroud of Turin. I will discuss the evidence surrounding the image on this cloth, which of all historical artifacts is the one that has been the most intensively scrutinized by scientists. I will present a case that understanding how this image formed might give us an insight into the relationship between mind and matter and thereby enrich our understanding of the nature and potential of humanity itself. Interest in the Turin Shroud waned after the famous 1980s carbon dating that seemed to suggest that it was medieval in origin. But, as I will show, nobody can account for how the image could have been fabricated. Also the carbon dating has been called into question, as it seems that the corner from which the sample was taken was highly contaminated by much more recent material that had been added to the cloth in a repair in the sixteenth century.

The second is the phenomenon of near-death experiences. These are episodes when consciousness continues while brain waves are flat, and the subjects are aware of their immediate and remote surroundings at a time when the brain is not functioning.

Third, there is the question of the role of the conscious observer in quantum mechanics in "making reality real." We will explore how this might work and consider the implications for understanding ourselves and each other.

You will see though that this is not a book about these three phenomena but instead is one about you, the reader. It is an exploration of the "I" that looks out at the world from "behind your eyes." The man of the Shroud was a human being, as are you. All his achievements lie within the realm of human potential and therefore your potential. In this book, we will consider evidence from the Shroud. We will also consider evidence from physics and from your human experience of consciousness, awareness, and free will that will show that *your* capabilities might be limitless and eternal.

Two of the themes of this book are the demonstration, through scientific evidence and reason, of the limitless potential of humanity and an exploration of how we have come to exist as individualized sentient beings. These themes will lead to an exploration of that which confines us and how such restriction might be undone.

The Intriguing Cloth

*What Is the Shroud of Turin, and Why Is Its
Image Still a Mystery?*

Of all the world's historical artifacts, the Shroud of Turin has been one of the most closely studied by science. Countless thousands of scientific and empirical man-hours (and woman-hours) have been invested into trying to unravel its mysteries. The fascinating result of all of this is that we still do not know a way, even with twenty-first-century technology, to produce an image that has the key and unique properties of the one on the Turin Shroud. In some of the chapters that follow, I will invite you to join me in exploring these unique properties and will discuss various ideas about how the image occurred. We will need to consider the historical evidence regarding where and when the Shroud appeared and also look at scientific evidence regarding the origin of the image and forensic evidence that suggests that it once wrapped the dead body of a crucified man. Reason and logic together with the methods of science are able to explain many phenomena that were once considered insoluble mysteries. When applying these methods to the Turin Shroud, can we perhaps find clues as to where the answers will lie? I will allow you to decide the answer to that for yourself if you will join me in this fascinating investigation through the pages of this book.

The aim goes beyond an understanding of the image on this fascinating piece of cloth. I will present startling evidence to you that suggests that there is an explanation for the image that illuminates far more than just this one subject. It may also shed some light on some of the most enduring mysteries of mankind, such as: How does consciousness exist? Is there such a thing as free will, and if so, how might it have come to exist? What is mind, what is matter, and how are the two related? You will see that the event that may have caused the image formation on the linen ties in with and supports the ideas presented in this book. However, the book does not rely on the Shroud for clarification or verification of these ideas. We will explore other streams of empirical evidence that all point to the same conclusions—conclusions that may invite us to reconsider what we are as human beings and what our full potential might be.

I first became familiar with the Shroud of Turin as a schoolboy in the early 1980s when a friend showed me a recording of David Rolfe's 1978 BAFTA (British Academy of Film and Television Arts) award-winning documentary *The Silent Witness*. The film presented strong evidence that the Shroud once wrapped the dead body of a crucified man and that the image on the cloth was not the work of an artist or a forger. Ever since then, I have been intrigued by this subject, and I have been conducting research into it for over thirty years. I have presented my findings at international scientific conferences, and my papers have been published in peer-reviewed publications.

How then did the image get onto the cloth? If it is not a work of art or a forgery, then what is it? I went on to study science and medicine, and the more I learned about the scientific research on the Shroud, the more fascinated I became by it. It is singularly the most studied historical artifact of all time, and still to this day nobody knows how anyone could produce an image with the characteristics of the Shroud image, even using modern space-age technology.

I have never been content with words like *mystery* or *miracle*. For me, these terms fall short of an explanation, and as the Shroud, with its bloodstains and image, can be studied scientifically and empirically, I would always prefer to seek a scientific explanation for them.

I have gradually come to the conclusion that this could be possible but that we might need to follow the clues in science that suggest that science is not yet complete. Quantum theory, for example, places consciousness at the heart of reality, and yet science does not account for how consciousness arises. Some eminent scientists such as Erwin Schrödinger would argue that consciousness did not begin but must be rooted beyond time itself.

When I was at medical school, I was as interested in physics as I was in medicine and would often attend physics lectures. While at university (1985–1991), I was privileged to attend a lecture given by Professor Sir Hermann Bondi on special relativity and had a chance to discuss with him one of the ideas about light that are presented in this book (see pages 70–71).

Professor Robert L. Morris, who was appointed as the first Koestler Professor of Parapsychology at the University of Edinburgh, also visited my university to give a public inaugural lecture. He was analyzing empirical evidence that investigated the possibility of the existence of extrasensory perception and mind over matter. During the Q and A session after his lecture, I raised my hand. When the chairman indicated that it was my turn to speak, I put it to Professor Morris that the act of someone deliberately raising his hand could be seen as evidence of mind over matter if we were to consider the existence of "free will."

You will see that these themes of "light" and "free will" figure prominently in the ideas that are presented in this book.

The event that formed the image on the Shroud is, as far as we know, unique in human history. I would like to take you with me on a journey to consider the implications of empirical observations about the Shroud image for our understanding of the cosmos and of humanity. Since it was a human being whose image was formed, I will discuss some possible implications concerning the nature and potential of all human beings.

In 2014, I presented a paper at a conference in St. Louis, Missouri. The title of my paper was "The Image on the Shroud: Natural, Manufactured, Miracle, or Something Else?" The full paper is available as a pdf file from the Shroud of Turin Education and Research Association at their Shroud of Turin website.[4] The main point I was trying to convey was that there seems to be a tendency for Shroud research to be split between those who believe the image was formed in a "natural" manner according to recognized physical and chemical processes that could in principle be reproduced through empirical research and those who believe that the image-formation process was "miraculous." My personal position is a third one. I suggest that the discoveries of twentieth- and twenty-first-century physics have taken us to a point where we now need a new paradigm for understanding the nature of consciousness and the relationship between mind and matter.

This could completely revolutionize our understanding of what we are as human beings and what our potential is. The existence of the Shroud image may well be an aid to a cogent elucidation of this in a rational manner without needing recourse to dogma or faith. In this book, I will present evidence from this and other sources to demonstrate that you, in common with each and every other human being, are greater in significance and power than the entire inanimate physical universe.

THE INTRIGUING CLOTH

The Turin Shroud is a fourteen-foot-long piece of cloth kept in Turin, Italy, that has some very unusual markings on it. These consist of burn marks, watermarks, bloodstains, and the faint image of the front and back of a man's body (see plate 1). Until around 1898, this seemed to be just an indistinct pattern that when viewed together with the bloodstains suggested that it was the image of a man who had been whipped and tortured and who wore a cap made up of numerous sharp objects. The evidence also suggests that he had been crucified in a manner consistent with the Roman method of crucifixion, as confirmed by other archaeological discoveries in the twentieth century. In 1898, Secondo Pia, an Italian lawyer and photographer, was allowed access to the Shroud for photography by the Savoy family, who were its owners. Photography was still a relatively recent development then, and this was the first time that the Shroud had ever been photographed. It is said that Pia nearly dropped his photographic plate in shock when he saw the negative that emerged from the photographs. The negative photo of the Shroud looked to all intents and purposes like a very clear "positive" photograph, which implied that the Shroud image itself had some of the characteristics of a photographic negative (see plates 2 and 4). This was a catalyst for the spread of scientific interest in the Shroud, as scientists wondered how such an ancient image could appear to be like a photograph. The clarity and resolution of Pia's image, obtained as a photographic negative of the Shroud, was such that many people at the time found it difficult to believe that it could have been genuine, and he was accused of performing a hoax. However, the photographic negative properties of the Shroud image were again demonstrated the next time that the Shroud was available for photography in the 1930s.

How, though, did the world's first "photograph" occur without a camera and without a lens or any other technological device? As you will see, the answer to this question points to some revolutionary and astounding implications.

In 1976, John Jackson, an American physics professor, discovered that the Shroud image had certain unique properties (as described below). He was fascinated by the possibility of using the scientific method in an attempt to explore how the image formed. Together with some other scientists, including many from NASA and the Los Alamos National Laboratory, he founded the Shroud of Turin Research Project (STuRP).

In 1978, STuRP scientists were given 120 hours of round-the-clock access to the Shroud for study using state-of-the-art equipment. This permission was granted by the Duke of Savoy, who was the owner of the Shroud. Each of the members of the team approached it from his or her own particular area of expertise. The brief of STuRP was to explain how the image could have been formed on the cloth. Many of the scientists went to Turin expecting to find evidence of paint or pigment and to "see the artist's brushstrokes" on the cloth. Instead, they found nothing of the sort, and they couldn't see how such an image could have been fabricated. The members of STuRP came from a great variety of backgrounds, and many of them were initially skeptical about the provenance of the Shroud, and yet after their study, they were all convinced that the Shroud image could not have been the work of a medieval forger or artist.

If the image on the Shroud had been painted, they would have seen an outline to the image, and they would have seen "directionality" caused by brushstrokes. Most importantly, when they studied the cloth microscopically and chemically, they would have found that the fibers would have been matted together, and the image would have been chemically composed of extraneously applied substances. They would also have found that some of the paint had soaked right through the cloth. None of this was found, and consequently, the scientists concluded that this was not a painted image.

They studied the chemical nature of the image and found that nothing had been added to the cloth to form it. The fibers that are woven together to form the linen can be viewed at the microscopic level, and when this is done, it can be seen that each fiber is composed of numerous fibrils. An oxidation and dehydration of some of these fibrils had somehow occurred, causing the color to change to a more sepia hue than in the adjacent fibrils (see plate 5).[5,6] This alteration in color seems to be similar to the effect you see when the pages of a book have been exposed to sunlight. This process in paper is driven by the ultraviolet rays in sunlight. As you will see, the evidence suggests that it may have been ultraviolet light that caused the Shroud image also, and, as a consequence, the thickness of the image is less than a thousandth of a millimeter. Strange as it may seem, the scientific evidence actually suggests that the source of the radiation that formed the image was not from the sun but from the dead body that was contained within the Shroud!

The STuRP team found that the variation of intensity of the image wasn't caused by how discolored the fibrils were but by the number of fibrils that had been discolored. The way that the image is contained on the cloth is similar to the way a black-and-white newspaper photograph is composed of numerous tiny pixels. There is no variation in the intensity of the coloration of the pixels, but instead the variation in intensity of the image is caused by the number of pixels per unit area. One cannot even see the image from close up but has to stand around two meters away in order to see it, and so if one postulates a forger being responsible for the image, then he or she would have to have in some way altered the cloth at the microscopic level, fibril by fibril, without being able to see the work while it was being done. Also the forger would have to have anticipated forensic and technological advancements that would occur several centuries later and have deliberately included information in his or her handiwork that would only become apparent, understood, and testable many hundreds of years later.

Once seen in a photographic negative format, the Shroud image suddenly switches from a vague, indistinct, apparent stain on the cloth to a well-resolved, clear and accurate picture of a man, from both the front and back (see plates 1–4).

If an artist or forger were able to produce such an image without any trace on the cloth of how it had been done, why produce an image that only becomes apparent once it is studied with technology that was undreamt of in that day?

STuRP was founded after John Jackson had a "eureka" moment similar to Pia's when he subjected the Shroud image to a device called a VP-8 image intensifier, which was a piece of technology usually used for interpreting X-ray images. What the VP-8 does is convert image intensity into three-dimensional relief. Jackson found that this process caused the man of the Shroud to appear to rise out of the cloth so that the three-dimensional contours of his body and face, both front and back, became visible. If one analyzes a painting or a photograph (positive or negative) of a person with a VP-8, there is no such three-dimensional rendering of the image. Instead, one finds an amorphous set of peaks and troughs that have no particular resemblance to the actual three-dimensional contours of the subject who has been photographed.

The notion that the Shroud image could have been painted onto the Shroud is not sustainable for several reasons, including these:

- The image is not visible from close to the cloth and is composed of altered coloration of parts of individual fibrils in a pixelated manner. If the "pixels" had been applied by a painter, then it would seem that he would have needed a seven-foot-long paintbrush and that the filament that applied the paint would have had to have been thinner than a human hair.
- The artist would have had to have painted the image one pixel at a time so that there were no brushstrokes, no directionality in the image, and no outline. He would have had to have painted the image in a photographic negative format while at the same time including "distance coding"* into the pixel density. He would have had to have achieved all this without leaving chemical traces of the paint in the image and without the paint soaking into the cloth or even causing a bonding together of the surface fibrils.
- Chemical and radiographic examinations of the Shroud image area have confirmed that the image does not consist of a chemical coating but rather of oxidation and dehydration of the cellulose in the linen.

Given all the evidence that the Shroud image could not have been fabricated, it is perhaps not surprising that physicist Paolo Di Lazzaro made the following statement in a newspaper interview:

*Distance coding describes the phenomenon that the intensity of the image depends on the distance there would have been between the cloth and the body, so that, for example, parts of the body that would have been closer to the cloth, such as the nose and the hands, have a higher intensity of image formation than parts that would have been farther away.

In summary, we cannot imagine a medieval forger who has a micro-scope and technology better than today's technology. We have to believe in a miracle to think that the Shroud is a medieval hoax.[7]

The other striking feature of the Turin Shroud is what appears to be bloodstains (see plate 6). These were chemically tested and were found to consist of human blood[8] and to have properties that contrast strongly with those of the body image, namely:

- Unlike the body image, the bloodstains penetrate through the thickness of the cloth.
- When they are viewed microscopically, one can see that, unlike the case of the fibers bearing the body image, the fibers bearing the bloodstains are indeed matted together.
- Where the bloodstains are present in the image areas, the image is missing underneath the bloodstains.

The first two points are not surprising and are what we would expect from bloodstains. The third point is particularly interesting as it implies both that the bloodstains must have occurred before the body image was formed and also that whatever the process of image formation was, it was precluded from happening through the bloodstains. As we shall see, forensic evidence suggests that the man of the Shroud was already dead when his body was wrapped in the Shroud, and so if the image formed after the bloodstains, then it must clearly have formed after he had died.

A CSI-STYLE FORENSIC ANALYSIS OF THE SHROUD OF TURIN

Remarkably, a crime scene investigation–style forensic study of the Shroud yields much detailed information about how the man whose body it once wrapped was tortured and killed.

The bloodstains and the body image are sufficiently detailed to allow us to conduct a forensic investigation of the Turin Shroud to derive information about the injuries that occurred to the man of the Shroud. What we see is that the Shroud once wrapped the dead body of a man who had been brutally tortured. There is evidence of tearing of the flesh in more than 120 places with wounds that have the pattern characteristic of those inflicted by a Roman flagrum. One can even deduce that there were two assailants, one on either side, based on the pattern of the injuries apparently caused by flagellation and the orientation of the lacerations to the skin. It is also evident that there was a significant difference in height between the two people who were whipping him. Between the shoulders, one can see a long area of contusion on the skin consistent with him having carried a heavy object such as a wooden beam. There are signs of injuries sustained in a fall, possibly while carrying this beam, with bruising around one cheek and swelling around the nose, as well as signs of a severe laceration to the left knee. It would seem that the left knee gave way and that his weight was mainly taken both by the left knee and by the face, which hit the ground, implying perhaps that he could not break his fall with his hands as these may have been tied to the beam.

A chemical analysis of the surface of the cloth demonstrated localized areas of dust (dirt) around the feet, the left knee, and the face. There is some preliminary evidence that suggests that this dust has a mineral composition consistent specifically with Jerusalem limestone.[9] I do have some reservations about whether the study that looked at the provenance of the dust was sufficiently rigorous and used appropriate controls—hence the description of it as preliminary evidence. The STuRP research, however, was performed in a neutral and rigorous fashion, and the findings were published in peer-reviewed scientific journals. The presence of the stone dust on the cloth only became apparent using twentieth-century technology, so it would not have been known about in medieval times, and therefore there would have been no need for a forger to have added it to the cloth to convince anyone of its authenticity. There are bloodstains over the forehead and all over the scalp, including the vertex of the head, implying that a "cap" of thorns may have been placed on his head. This is particularly interesting as medieval art typically shows merely a circlet of thorns, but from this evidence, it would seem more likely that an amorphous clump would have been placed on his head.

There are bloodstains visible around the left wrist. The hands are crossed so that the right wrist is not visible. The bleeding from the wrist is heavy, and two streams of blood are seen as though blood had flowed down the forearm toward the elbow. If we consider which position the arms would have been in to make these streams close to vertical, it is implied that the hands were raised above the shoulders as though the body had been suspended by the wrists, as in crucifixion. Looking at the bloodstains on the feet, it appears that the feet also had been impaled, and this is consistent with another archaeological finding of a bone from a victim of Roman crucifixion in Judaea within which a nail is still present, which suggests that this may have been part of the practice of crucifixion.

There is also a wound visible on the lower chest from which issued blood mixed with clear fluid (possibly pleural fluid). This is consistent with the report of a man being stabbed by a lance.

Robert Bucklin, who was the forensic examiner in Los Angeles in the late 1970s, said after studying the markings on the Shroud:

> The markings on this image are so clear and so medically accurate that the pathological facts they reflect concerning the suffering and death of the man depicted here are, in my opinion, beyond dispute.[10]

John Robinson was the dean of Trinity College at the University of Cambridge. He wrote books such as *Honest to God,* which earned him a reputation for having a very skeptical approach. He assumed that like all other "relics" he had encountered, the Shroud would give him the impression of being bogus.

However, the more Robinson found out about the Shroud, the more he realized that there were many features that a forger would never have thought of. When interviewed for the documentary *The Silent Witness,* he said, "In fact what we have fits extraordinarily well with the New Testament evidence and helps to make a great deal more sense of it than I saw before."[10] The New Testament would not have told a forger that the nails of crucifixion would have been placed through the wrists. The weight of the body, if carried by the hands rather than the wrists, would have caused the flesh to tear.

Swiss criminologist and pollen expert Max Frei studied the pollen on the Shroud and concluded that it had spent time in Europe but also around Constantinople and the Middle East and in particular around the area of Jerusalem. This has also been confirmed more recently by research done by Professor Avinoam Danin of the Hebrew University of Jerusalem.[11]

Historian Ian Wilson made a study of historical depictions of Jesus and found that before approximately the sixth century there was very little consistency in the depictions but that after that time they all tend to be similar to the Shroud image. This is interesting, as around that time a cloth is reported to have been discovered—or perhaps "rediscovered"—in Edessa (now known as Şanlıurfa) with a mysterious imprint of Jesus that was described as "not made by human hands." As noted, the pollen evidence does suggest that the Turin Shroud may well have been around this part of Turkey at some point in its history. Legend has it that the historical King Abgar V, who was king of Edessa in the time of Jesus, was suffering from an illness and, having heard of Jesus' famed power to heal people, had sent him a message asking him to come to Edessa to cure him. The story goes that Jesus did not himself come during his lifetime but that instead one of his disciples, Addai (perhaps Thaddeus?) came with a cloth bearing a "miraculous" image of Jesus on it.[12] Strong evidence suggests that this is the same cloth that is now in Turin. For more detailed discussion of the historical evidence, I would direct you to the work of historians such as Ian Wilson and Dan Scavone, professor emeritus in ancient and medieval history at the University of Southern Indiana.

According to Jewish burial practice, the blood that is shed at the time of death must be buried with the body.[13] Under normal circumstances, the body should be washed before burial. However, if blood shed during life is mixed with blood shed at the time of death, then this is referred to as *mingled* blood, and this blood is supposed to be buried with the body, and so the body would not be washed. This is noteworthy, as it appears from the evident bloodstains on the Shroud that in this instance the body does not appear to have been washed before it was placed in the Shroud. It was the American physician Gilbert Lavoie who drew the world's attention to this in the context of the Shroud.[14] At this point, a quote from the ancient Jewish text the *Mishnah* is relevant.

> What counts as "mingled blood"? If beneath a man who is crucified, whose blood gushes out, there was found a quarter-log of blood, it is unclean; but if beneath a corpse whose blood drips out, there was found a quarter-log of blood, this is clean.[13]

A "log" of fluid is defined in an appendix to the *Mishnah* as a volume equivalent to the contents of six eggs, and so a quarter-log would be equivalent to about one and a half eggs, or about a small cupful of blood. The word *unclean* is used for blood shed at the time of death; it is considered as *blood of atonement* for sins according to Jewish law.

According to Roman law, a crucified person remained on the cross as food for birds of prey.[15] The practice of burial for a crucifixion victim was confined to only a small region of the empire and only for a very short period. In 6 CE, the emperor Augustus removed the Jewish king Archelaus (son of Herod I) and installed a Roman procurator for Judaea and Samaria who had the authority of the death sentence. At the same time, however, a Jewish government was still in existence, which required burial before sunset according to Jewish law. This exceptional double rule was finished by a war in 66 CE known as the Jewish revolt against the Romans.[16] Consequently, we have evidence consistent with a date for the crucifixion of the man on the shroud between 6 CE and 66 CE.[17] This is because the bloodstains on the Shroud demonstrate that the body was placed in the Shroud soon after death occurred, as otherwise the blood would have dried to the extent that there would have been no staining of the cloth when it wrapped the body. Also, there is no evidence of putrefaction on the body image on the Shroud.

In Oviedo, Spain, there is another cloth known as the Sudarium. The provenance of this cloth is well documented, back to when it was brought to Spain from Jerusalem around the year 400 CE. The Sudarium of Oviedo is also marked by bloodstains. The interesting thing is that the pattern of the bloodstains matches extremely well with the bloodstains around the face and head of the man on the Shroud. Also, there is evidence that the blood group of both stains is AB (an uncommon blood group).[18] Forensic experts have concluded from all of this that there is substantial evidence that the Sudarium and the Shroud once wrapped the same corpse.

It is conceivable that before the man was wrapped in the Shroud, a cloth might have been wrapped around his head to soak up some of the blood and that the face cloth might then have been placed in the tomb separate from the body in the Shroud. There would seem to be some interesting evidence here regarding the provenance and age of the Shroud in that it would seem that the two cloths could only have coincided in location in or before the fifth century, and it would seem that the most obvious candidate for where they would have coincided would be Jerusalem, as this is where the Sudarium was until it was brought to Spain.[19]

So we have seen that there is substantial evidence that the Shroud is far older than the carbon dating suggested and also that the image could not be manufactured, even with modern technology. If it could not have been manufactured, then how could it have formed? If you read on, you will see that I will suggest a rational explanation that also addresses the question of what we are as human beings and why this means that consciousness could never be manufactured in a machine.

How Did the Image Form?

As mentioned in the first chapter, the scientific study of the cloth currently known as the Turin Shroud made great progress in 1978 when the Shroud of Turin Research project (or STuRP) was allowed 120 hours of round-the-clock access to the Shroud over five days and nights. The researchers' brief was to scrutinize the image using state-of-the-art technology and their own scientific expertise and to explain how the image formed. The team included four photographers and twenty scientists, including, for example, two from the U.S. Air Force Weapons Laboratories, two from the U.S. Air Force Academy, two from NASA's Jet Propulsion Laboratory, and five from the Los Alamos National Laboratory. (To see pictures of the scientists examining the shroud, see plates 7 through 10.)

From their extensive microscopic, radiographic, and chemical analysis, they concluded that they could not find a method by which the image with all its unique characteristics could have been manufactured, even with twentieth-century technology. To date, nobody has been able to account for how an image with these properties might be manufactured with twenty-first-century technology either! After the carbon dating in 1988, many people have assumed that the Shroud image must have been manufactured in medieval times. However, there is substantial evidence that suggests that the sample that was removed for the radiocarbon laboratories (which did not bear a part of the image) was mostly composed of much more recent material that had been used to repair a damaged corner of the cloth in the sixteenth century, thereby explaining the erroneous results!

For those of you who are interested in a slightly more technical summary of the carbon-dating issue and some of the detailed findings of STuRP, I have included these in the appendix on page 167.

The chemical changes in the image-bearing fibrils are the same as those in the cellulose of paper, which turns sepia yellow upon exposure to sunlight. Could this be a clue to the mechanism of the changes that occurred in the fibrils to cause the image to form?

Research at ENEA has demonstrated that certain key features of the image can be replicated by projecting short-duration high-intensity ultraviolet rays on the surface of linen.[20] Linen is not transparent to short-wavelength ultraviolet light, which is absorbed by the outermost surface layer of the cloth. Di Lazzaro and his colleagues demonstrated that in certain conditions ultraviolet irradiation can produce the same chemical change in the linen as is found in the image-bearing fibrils of the Shroud and also that this change is limited to only these outermost superficial fibrils.

Therefore, the extreme "thinness" (superficiality) of the Shroud image is in keeping with an ultraviolet energy source being responsible for the chemical change in the surface cellulose. How does this relate to two other key features of the image mentioned in chapter 1: the photographic negative properties and the three-dimensional distance-coded information it contains?

The photographic negative properties of the image (see plates 2 and 4) suggest that the image may well have been caused by radiation of some sort, and the fact that there is distance-coded information contained in the image implies that the radiation appears to have emanated from the dead body of the man who was wrapped in the Shroud. As we have seen, research at ENEA suggests that the nature of the image is consistent with ultraviolet irradiation having been the cause of the image.[20] This possibility was originally suggested by physicist John Jackson of STuRP in the 1970s.

Di Lazzaro and colleagues do point out that they are not suggesting that the image was formed by a technological process. On the contrary, this could not be for the simple reason that their calculations demonstrate that producing the whole Shroud image (if that were possible) in a similar way would require a thirty-four-thousand-billion–watt laser source, which is not available even today.[21]

Just for comparison, it is worth bearing in mind that the most productive of nuclear power stations, for example, are able to produce a peak power output of around four billion watts, and the wattage of the process that formed the image on the cloth appears to have been over a thousand times more than this! For those not familiar with the physics definitions of power and energy, I should clarify the distinction. Power (watts) relates to the *rate* of transfer of energy. That is why energy consumption in our homes is measured in kilowatt-hours. As the suggested burst of light from the man of the Shroud would, according to the experiments at ENEA, have been for only a tiny duration of time, the total transfer of *energy* to the cloth would not have been enormous. (If it had been, the whole cloth would have been destroyed rather than changed by the formation of an image.) Nevertheless, it is still remarkable that no synthetic source of coherent light of this magnitude of power is known to us today. This of course adds more weight to the cumulative evidence that the Shroud image could not have been manufactured or formed by a technological process, especially one that people might suggest occurred many centuries ago.

Let's just take a "time out" here and consider this astounding implication again. There is scientific evidence that suggests that the dead body of a crucified man many centuries ago emitted a sudden, short, intense burst of radiant energy in a manner that we cannot replicate even with twenty-first-century technology!

There is an obvious question which arises from all this:

If we are to believe the evidence that the Turin Shroud image was caused by a momentary, intense burst of ultraviolet light from the dead body of the man whom the Shroud wrapped, then what might the cause have been for this burst of radiant energy?

In the pages to follow, we will consider a way of answering this question.

Let me assure you at the outset that the ideas suggested do not rely on invoking "mystery," "miracle," or "magic." The Turin Shroud is a physical object that we can study using the empirical tools of science, and using only these tools, its remarkable nature is revealed.

The image on the Shroud has been studied in great detail using photographic, chemical, and microscopic methods. This cloth wrapped the body of a man who had been tortured by being whipped numerous times. He had had what appears to have been a cap of thorns impaled on his scalp and had been made to carry a heavy beam across his shoulders before being crucified with nails driven through his wrists and heels. He had died on a cross and then his body had been taken down and wrapped in the Shroud.

WHO IS THE MAN OF THE SHROUD?

We have seen that the empirical evidence regarding the Shroud suggests that it once wrapped the body of a recently deceased man. This was someone who had been tortured by being whipped numerous times and then crucified. He would also appear to have had a cap of thorns placed on his head prior to his death, and all of this would seem, according to empirical evidence from pollen found on the Shroud and other data, to have happened in either March or April in the environs of Jerusalem at some point between the years 6 and 66 CE.[11,17]

As the "crown of thorns" was a very specific form of mockery and torture that is said to have been devised specifically for the individual known as Jesus of Nazareth, this together with all the other evidence mentioned above strongly suggests that the man whose image we see on the Shroud is none other than that of the same Jesus of Nazareth.

As far as we know, the formation of the image on the Shroud was a unique event in human history. Even now, with twenty-first-century technology, we do not know how to duplicate an image with the properties described above (and in the appendix).

The man known as Jesus of Nazareth also arguably had a unique impact on the world through what he taught and how he lived.

Is it merely a coincidence that this unique event in human history (the formation of the Shroud image) appears to be associated with this particular individual?

If it is not coincidental, then perhaps there might be clues in his teachings and life that might help us to understand how the image formed.

It is a historical fact that people have formed religions that purport to be based on the teachings and life of this particular individual, but the Shroud image formed before these institutions did, and therefore it could be a piece of direct evidence to be considered independently. As such, perhaps it affords us an opportunity to understand Jesus in a fresh way. It connects us directly to his personal journey, thereby bypassing two thousand years of history and the politics of religious institutions.

The Shroud image is an empirically verifiable phenomenon. It does provide strong supportive evidence regarding the historical existence of Jesus of Nazareth, and indeed, there is strong evidence that the Shroud may well have been his burial cloth. As such, perhaps it affords us an opportunity to understand him in a fresh way without basing this on religion, theology, or dogma.

As he was a human being, I will show that an understanding of him (and of the image-formation process) may give us some clues to help us understand the nature of humanity itself and our relation to each other, as well as our relation to the physical universe.

It's a big claim, I know, to suggest that an old piece of cloth is so significant. However, it might have been a "silent witness" to an event that is unique in human history and might even hold the clues to the answers to the big questions that have puzzled philosophers and thinkers through the ages—answers written not in religious edicts or dogmas but in simple reason.

Others who try to account for the image of the man on the Shroud tend to do so in one of three ways:

1. Some people believe that the image was fabricated, perhaps by a medieval artist. I would argue that since the key properties of the image cannot be replicated, even using twenty-first-century technology, it would seem highly unlikely that a medieval forger could have achieved this. Di Lazzaro in a recent interview put it very well when he said, "*We have to believe in a miracle to think that the Shroud is a medieval hoax.*"[7]

2. Another theory of image formation is that the natural release of chemicals (amines from amino acids) from the decomposing body that was wrapped in the Shroud caused a chemical discoloration at the surface of the cloth via something known by chemists as a Maillard reaction, which is a similar reaction to the one whereby peeled apples turn brown when exposed to air. However, there are several problems with this. Maillard reactions have never been known to produce high-resolution images in other contexts. As far as we know, the Shroud image is unique, but if the image formed according to normal chemical processes, then why are there not many more similar examples? Also, decomposition products would always tend to be more concentrated at the orifices

of a body, and yet we do not see any distortion or increased intensity of the image, for example, at the mouth and nostrils. And we would not normally expect much amine production from inert structures such as hair, and yet these are also imaged on the cloth.

3. There are others who would argue that the Shroud image is "miraculous" in nature and therefore cannot be understood by recourse to natural law. However, the Shroud is a physical object that can be and has been studied empirically. The dead body of the man who was once wrapped by the Shroud was also a physical object, even to the extent that his wounds can be empirically studied from a forensic point of view based, for example, on the pattern of bloodstains on the Shroud.

If, as the evidence from the ENEA studies suggests,[20] there may have been a momentary, highly intense burst of ultraviolet light that emanated or "shone" from the body that the Shroud wrapped, then this would also have been a physical event that produced physical effects on the cloth that can now be studied scientifically.

This of course leaves some questions open:

What might have caused a dead body to produce such a momentary burst of radiant energy, and what might be the underlying mechanism for this?

Once we start appealing to notions of a "miracle," then anything goes. I would therefore agree with those scientists who try where possible to seek an understanding of all phenomena via natural law.

In view of the correlation between the unique teachings and life of Jesus of Nazareth and the unique characteristics of the Shroud image, perhaps there is another option for seeking an understanding of the mechanism of image formation. Perhaps we need to extend science's understanding of natural law to include the mind and its connection to matter.

In the chapters that follow, I will suggest where the clues might lie to how such an understanding could begin. I will present the case in this book that the power of the man who left the image on the Shroud lay in his humanity! We all have the power of mind over matter, but this may be diluted if we think and behave in ways that are less than human. We are not machines, but if we see other people merely as objects and treat them as such without the highest degree of empathy, then our behavior is machinelike and hence the power of our humanity is diminished.

So, to summarize, the changes in the surface fibrils that constitute the image are physically present on the cloth, and they can be studied empirically and scientifically. Empirical evidence suggests that the image may have occurred as a result of a momentary burst of radiant energy that emanated from the corpse that once was wrapped in the Shroud. Is this possible, and if so, can we use a rational, scientific approach to enable us to understand how this might have happened?

A PHOTOGRAPH OF A UNIQUE EVENT

As explained in chapter 1, it is an astonishing fact that the image of the man on the Turin Shroud has all the properties of a photographic negative rather than of a painting. Not only that but the distance-coding properties of the image also suggest that the "light source" in this case was none other than the dead body that the Shroud once wrapped. I am aware of the immensity of this claim, but this is the inference that we are led to by simply following the evidence collected from scientific studies of the cloth in conjunction with the results of the ENEA experiments. This dead body may well have shone momentarily brighter than the sun!

The evidence presented in this book could indeed imply that the image on the Turin Shroud might have occurred as a result of a momentary, intense burst of radiant energy that emanated from the dead body that was wrapped in the cloth. The experiments carried out by ENEA in Italy confirm that the characteristics of the Shroud image indicate that this could be how it happened. If the image did appear on the cloth in this way, then how could we attempt to explain this?

Could there be an explanation for the image-formation process that is neither technological nor miraculous? To address the question of whether the image on the shroud is natural or miraculous, we would need to have a working definition of what we mean by the word *nature*. The notion of a miracle is often taken to refer to something that is supernatural or outside natural law.

For most scientists, nature is considered to be the set of all observable phenomena. The term *observation* is generally used to suggest experience acquired directly or indirectly via the senses, such as sight and hearing. The adjective *empirical* is used by scientists to relate to evidence that is seen or experienced via the senses. Dictionary definitions of the word *empirical* tend to refer to experience rather than just specifically sensory data, with terms such as "based on," "concerned with," or "verifiable by observation or experience rather than theory or pure logic."

As technological devices cannot see or measure a mind, the empirical view of the world tends to be one of electromagnetic waves or vibrations in the air rather than one of color or sound, for example. You cannot break down the contents of your experience into terms that can be objectively described.

If you gaze upon a red rose and inhale its fragrance while experiencing the pleasure that the flower might give you and reminisce about memories associated with it, a scientist could explain to you that its redness is the result of the fact that the light reaching your eyes has a wavelength of around seven hundred nanometers, which is seven ten-thousandths of a millimeter. That *is* the wavelength of the light and you *do* experience the rose as red, but this explanation doesn't account for where your experience of "redness" comes from. Similarly, its fragrance is accounted for as the result of nerve cells in your nose being stimulated by particular shapes of molecules produced by the rose, but again, many neurobiologists will concede that this does not explain your experience of the fragrance of the rose. It merely describes how your nose detects the molecules that are associated with that fragrance. In fact, none of our sensations of each other or the world around us can ever be explained within the empirical view of the world.

So our perceptions themselves are ignored, but most importantly, the elephant in the room remains undeclared. The mind that perceives these experiences is itself invisible and silent and is therefore not a part of the empirical model of reality. But the empirical model of reality has no meaning unless minds exist, as its very definition depends on experience, and to experience one must be aware, that is, sentient!

As the empirical model of the world doesn't contain mind, it is understandable that people who view the world in that way use the term *dualist* to describe someone who suggests that mind might interact with matter. To many empiricists, accepting that such a thing might be possible would imply that a mysterious "something" from outside the empirical world reached into it and changed it. Yet, with the greatest irony, many such people seem unaware that this division of the world into two parts or categories, the observer and the observed, is implicit to the empirical worldview itself!

Erwin Schrödinger, the Nobel Prize–winning founder of quantum theory, felt that the empirical view of the world was empty and incomplete:

> We step with our own persons back into the part of an onlooker who does not belong to the world, which by this very procedure becomes an objective world.[2]

Colour and sound, heat and cold, are our immediate sensations. Small wonder that they are lacking in a world model from which we have removed our own mental person.[2]

The material world has only been constructed at the price of taking the self, that is, mind, out of it, removing it; mind is not part of it; obviously, therefore, it can neither act on it nor be acted on by any of its parts.[2]

I remember when I was just nineteen years old a friend pointed out something to me that made such an impression on me that I still think about it now, over thirty years later, so I'll share that with you also. Take a look up at the night sky (best done somewhere in the countryside where there is less light pollution) and see the countless billions of little white dots. Each of them is a star just like our own sun, and the huge size of what you can see in the night sky is almost inconceivably large. But this whole vista exists in you, in your mind's eye. Now consider that because you can see all that, your mind must be "bigger" than what you see, as this perception is contained within you. So compared with the enormity of the whole physical universe, a human being *seems* like a tiny little microscopic dot, yet, in reality, the entire physical universe is like a tiny dot compared with you. Now, next time you are out in a crowd, surrounded by people, just consider for a moment that each of those people is bigger than the whole physical universe. Mind-blowing, isn't it? If we realize this *enough,* then perhaps we might value ourselves more and also value everyone else more too! The amazing thing is that physics itself verifies this significance, as you will see in the following chapters.

I find the implications of this to be astounding. It demonstrates something about what we all are as human beings that gives each individual human being a significance greater than any object, be it a rock, a planet, a galaxy, or even the whole physical universe, as we can perceive the world but it cannot perceive us. I will elaborate on this further throughout this book. Next, let's consider a possible alternative way of understanding what we are as human beings from the point of view of the interaction between mind and matter.

Of Mind and Matter

Do you believe that you have the power of mind over matter? If your answer is no, then my next question would be, When considering your actions, do you believe that you are responsible for them in the sense that you accept the credit or blame for what you decide to do? We wouldn't generally blame a boulder for falling onto a road or award a Nobel Peace Prize to a computer (well, let's hope not, anyway!), so why do we feel that *we* are responsible for our actions?

Surely, it is because of the tacit recognition of "freedom of will." If we "could have done otherwise" than what we did and chose to do, then what we did was surely our responsibility. The boulder follows the laws of physics, and the computer follows its program, but it clearly seems to us as human beings that we can act freely. If that is the case and our minds make choices that influence what our bodies do, then how can you not believe that you have the power of mind over matter?

Many people assume that if there were a power of mind to influence matter, that this would be by definition *supernatural*. This is perhaps because it is often assumed that in nature, matter obeys physical laws that define what matter is and what it does. The mind is further assumed by many people itself to be a product of these same physical laws, and it is therefore often said that the relationship between mind and matter is merely one of mind being determined by matter. It is also therefore commonly assumed that matter cannot be influenced by the mind.

However, although these are common abstract assumptions, if we look at our notions of personal responsibility, accountability, morality, and so on, these all presuppose that we have free will. So, on the one hand, many of us believe that mind has no influence on matter, and yet in our day-to-day lives, we all live our lives based on the presumption that our minds are constantly influencing the matter of our bodies through the choices that we make by the exercise of free will. How would we define free will if it exists?

Will, according to a dictionary definition, implies determination by an act of choice. The designation "free" when applied to this suggests that there is no compulsion or force that compels one to make a specific selection from available options. This would mean that any sentient being that can exercise free will is able to harness what I would call a "primary cause." Material structures would not in and of themselves have the power to perform this simple act of choice. To quote Schrödinger again:

> My body functions as a pure mechanism according to the Laws of Nature. Yet I know, by incontrovertible direct experience, that I am directing its motions, of which I foresee the effects that may be fateful and all-important, in which case I feel and take full responsibility for them. The only possible inference from these two facts is, I think, that I—I in the widest meaning of the word, that is to say, every conscious mind that has ever said or felt "I"—am the person, if any, who controls the "motion of the atoms" according to the Laws of Nature.[2]

If we do have free will, then we are constantly shaping our actions and our nature through the choices that we make. All it would take is one person with free choice to mean that the world could not be completely predetermined or random, and there are several billion human beings!

Our sentience could *itself* be evidence indicating the freedom of our will. By sentience I mean the state of being aware. This does not depend on the content of that awareness but describes the fact that awareness exists. A simple analogy would be the word *space,* not in the sense of outer space specifically, but just space as in room for something to exist. Your table takes up space, but space would still be there if its contents were different. So sentience could be seen metaphorically as the "space" within which ideas, perceptions, experiences, emotions, and so on exist.

As you are reading these words, you are aware of the words and their meaning, but at a more fundamental level your awareness is bigger and far more significant than the words. The fact of your being aware, whatever you are aware of, demonstrates that you as a human being are greater in potential than the entire physical universe. Matter follows physical law, but as you have *awareness,* I would suggest that *you* can choose between options and therefore have free will. In other words, the choices you make are *not* determined by physical law. It could even be that the choices made by conscious beings such as you and me are the only things in the physical universe that are not determined by physical law!

If the implications of quantum mechanics about the nature of reality are considered, then physicality itself is a derivative of consciousness. Not only are you not determined by physical law, but the very existence of physicality and therefore physical law is only possible because sentient beings like you exist!

John Searle's "Chinese room" argument makes a strong case that there is a fundamental distinction between consciousness and the type of information processing that occurs in a computer.[22] Roger Penrose, emeritus professor of mathematics at the University of Oxford, in his book *The Emperor's New Mind,* presented good evidence suggesting that sentience cannot be created in the form of a simulation, for example, in a computer.[23] This would imply that there is no such thing as "artificial consciousness."

....

Consciousness Cannot Be Manufactured
The Chinese Room Argument

John Searle's Chinese room argument can be summarized as follows: Searle imagined himself (someone who does not understand the Chinese language) alone in a room, and pieces of paper with Chinese characters on them are passed under the door to him. He manipulates these characters according to an algorithm (like a computer program). Although he doesn't understand what any of the characters and numerals represent, he manipulates them just as a computer does to produce strings of Chinese characters as an "output" from the Chinese room that is a sufficiently appropriate response to the "input" that was delivered under the door to convince those outside that there is a Chinese speaker in the room.

Searle's point is that all he has done is manipulate information and he has not relied on any comprehension of what that information "means." The implication is that however sophisticated the output of a computer, it is always merely operating according to a program, which implies that it is never actually "aware" of anything. Early computers were more clearly mechanical in nature, and so it was readily apparent that they could not be conscious or sentient beings. Although computers have become more complex and more efficient at processing information, their essential nature has not changed. They are still machines just as a lever or an abacus is a machine. There is a deep logical flaw in the notion that one could manufacture an artificial mind.

Could Consciousness Ever Be a Product of Matter?

Many people who speculate on a physical basis for consciousness believe that it is what happens when a "system" can process information about itself and that through "self-reference" it also has self-awareness. Consciousness is thought to arise as a result of this. When I attend conferences about consciousness and present my ideas there, I often show a slide of a camcorder filming its reflection in a mirror to demonstrate what I see as the absurdity of this notion. I ask the other delegates there whether they think the camera is conscious as it is processing information about itself!

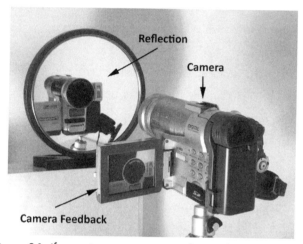

Figure 3.1. If consciousness were merely an object processing information about itself, then this camera would be conscious. Image courtesy of Daniel Langsman.

The neuropsychiatrist Giulio Tononi has postulated that consciousness arises out of the "integration of information," according to his integrated information theory of consciousness. For me, this misses the point. I would always contend that the point about consciousness is that it cannot be made of information, nor can it be a property of the *processing* of that information, as its fundamental nature relates instead to the capacity to be *aware* of information rather than to the *contents* of that awareness. Processing information merely rearranges or changes that information, but it does not invoke a *mind* that experiences the information. If it did, then that would mean my pocket calculator is thinking when it performs a calculation and that the camera filming its reflection is *experiencing* looking at itself! The fact that awareness is more than just the contents of perception was realized by many people at least two thousand years ago. In the *Kena Upanishad,* which is over two and a half thousand years old, one may find the following quote:

> That one which cannot be understood by the mind but because of which the mind is capable of understanding something is Brahman. Understand that alone to be Brahman. All other things that are being defined as "Brahman" and worshipped are not.

> That one which cannot be seen by the (naked) eyes but because of which the eyes are capable of seeing is Brahman. Understand that alone to be Brahman. All other things that are being defined as "Brahman" and worshipped are not.

> That one which cannot be heard by the ears but because of which the ears are capable of hearing is Brahman. Understand that alone to be Brahman. All other things that are being defined as "Brahman" and worshipped are not.[24]

As we shall see in the chapters that follow, quantum theory suggests that matter has no independent reality in the absence of a conscious observer. This being so, it would follow that consciousness cannot logically be dependent on matter for its existence. If we have awareness of options through consciousness, then this should be a clue to the reality of free will.

If we have free will, then that would mean that our actions are not entirely determined by any combination of nature and nurture but that there must be a third independent and undetermined factor. This would mean that a choice made with free will could be a cause that is not itself fully determined by external influences. In other words, it would be a form of primary causation.

If there is an effect, there must be a cause. If that cause is a result of another prior cause, then I would call that "secondary causation" or "dependent causation." That being so, we are led logically back to look for primary cause(s) to account for this. My contention is that not only does free will imply the existence of primary causation but that primary causation also implies the existence of free will. In other words, I am arguing that they are one and the same. I would suggest that this is self-evident in the same sense that "I think, therefore I am" is self-evident. Descartes wrote "Cogito ergo sum," that is, "I think, therefore I am." I would say, "I experience awareness, therefore I exist. I exist, therefore I can choose." While the nature of objects is open to subjective interpretation (e.g., Is this a messy bed or a work of art or both?), your existence as a sentient observer is known by you as an objective fact. After all, even if all your experience were an illusion (and I do not suggest that it is), then there still needs to be a "you" to *perceive* that illusion. It's ironic, isn't it, that if you think about it, the most objectively verifiable fact is that subjectivity exists?

If primary causation does exist, "where" would it function? If free will is the free choice between options, then is it not logical to assume that it would be found within a sentient being? Surely, one must be able to be aware of options existing in order to choose freely between them!

As I have said, I would go further and suggest that sentience implies the existence of free will, as once there is an awareness of options there is the scope to choose between them.

It is important to distinguish freedom of choice from freedom of choices. Someone in a prison cell may not necessarily be able to choose to be outside of the prison cell, but he is still able to choose between options of what he will do within the cell and also to use his will to move his mind's eye, for example, through choosing to focus on memories of the time before he was in the cell. Any free choice between options, whatever these options may be and however limited they are, is absolute freedom in the sense that it is a primary cause and in the sense that the chooser could have done otherwise. I am therefore suggesting that our very sentience as human beings is fundamental evidence that we do indeed have free will. If not for sentience, there would be no meaning behind the concept of "now" or "the present" as all points in time would be equivalent, without a "cursor" to highlight the point of "experience."

If, as Schrödinger implied, "the present" is made real as the present by us (any sentient observer), then it is not unreasonable to imagine that the future is undefined and can be described in terms of potentiality or possibilities for the very reason that through our choices we each have a role in shaping it!

In his books such as *What Is Life?* Schrödinger made some fascinating observations about the nature of mind, and consciousness in particular.

One of these was that the present—the "now"—has no explanation in known physical law but is made by the mind, which thereby creates time itself! His realization was that if time is the product of mind, then mind cannot be a product of time. Logically, therefore, it seemed to him that mind must be eternal, without beginning or end.

> I venture to call it [the mind] indestructible, since it has a peculiar time-table, namely mind is always *now*. . . . This means a liberation from the tyranny of old Chronos. What we in our minds construct ourselves [time] cannot, so I feel, have dictatorial power over our mind, neither the power of bringing it to the fore nor the power of annihilating it.[2]

If he was right, then that would mean that sentience is not ended by physical death. This has great relevance to our study of the Shroud, as empirical forensic evidence suggests that the body image on the Turin Shroud clearly appeared *after* the bodily death of the man whom it wrapped.[8]

So, to recap, the Nobel Prize–winning physicist Erwin Schrödinger made several fascinating deductions about the nature of mind and sentient awareness, three of which I have quoted.

Schrödinger's first point is that there is a tendency in science to look at the world exclusively in an empirical way. This method is based on the assumption that we can be surer about things or information if we base our statements about them on *measurements.*

Measurements do have advantages as means of gleaning information as they can be quantified, and they are also reproducible in the sense that others can repeat the measurements and check our findings.

Measurements often involve technical equipment, but we generally become aware of the results of measurements by receiving information via the senses. For example, a scientist might read an article in a scientific journal and become informed of the results of measurements in this way via the sense of sight. Also, he or she might be informed about the results in vocal conversation with a colleague and obtain information through the sense of hearing. Alternatively, scientists might themselves be the experimenters, in which case the information is still generally received via the senses.

However, the "mind's eye" is, itself, invisible. In other words, you cannot weigh a mind or see the mind that itself perceives or becomes aware of the data that emerge from your measurements. You can analyze the brain and measure the electrical activity that seems to be associated with sensory perception, but you cannot see the mind that perceives. Also, you can only see the "perception" by being the perceiver yourself. You do not generally see the perceptions of others directly.

So the first of these three points that Schrödinger made was that the empirical scientific worldview does not include the mind or perception. It is, therefore, a worldview in which we all as sentient beings (whether scientists or not) are absent.

Schrödinger himself was often careful to make a distinction between his philosophy and his discoveries in quantum theory. However, some other quantum physicists (such as the Nobel Prize–winning physicist Eugene Wigner) have suggested that whereas the empirical worldview ignores consciousness, quantum theory suggests that consciousness is fundamental to the very nature of reality. More on this later.

The second point was that a case can be made that our direct experience as sentient human beings supports the notion of free will, which, in turn, implies that mind can influence matter.

The third point is that time is itself a property of mind (see chapter 5) and as such, Schrödinger argued, mind must be eternal—without beginning or end.

The evidence from quantum mechanics suggests that mind might be a natural and essential foundation of the very existence of matter. That being so, why should we consider an influence of mind over matter to be supernatural? Perhaps it is natural for mind to be able to influence matter. We may perceive information about the world via the senses, but a physical object, whether it be a camera or a brain, does not of itself "see" the world. The concept of, for example, vision or hearing depends on the presence of a sentient being to *experience* these sense perceptions. This experience may sometimes have a counterpart in electrical patterns in the brain, but these patterns are just that, a *counterpart,* and are not the experience in themselves. This would be true in the same sense that if you walk on a beach, your footprints in the sand are a counterpart of your path along the beach.

If I am right, then all perception, whether or not physical senses are involved, is extrasensory perception. Also, if our actions are to any extent influenced by our choices made through free will, then a strong case could be made that they represent the effect of mind over matter.

To summarize, it is my contention that all our perceptions are extrasensory perceptions and that all our actions imply the power of mind over matter. With this in mind, I would suggest that the very word *supernatural* becomes redundant, as perception and action are completely natural to us as sentient beings!

MIND AND MATTER

Imagine, if you will, being in a state of unbounded potential where all knowledge that has ever been known or will ever be known exists within you and you can experience all of existence from all points of view from a perspective with no beginning or end. Now imagine having that same perspective now within your own mind.

Just as an exercise, try to imagine what it might be like to experience the world from behind the eyes of your nearest and dearest. Now imagine extending this to people you don't know and then to people you dislike. Finally, try to glimpse what it might be like actually to *be* awareness itself. This might sound difficult to do (some might say impossible), but I put it to you that you already *are* awareness and you *are* free will. The difference between us and the unbounded state is only a matter of degree and extent. The essential nature of being aware, including awareness of options and the ability to choose between them, is the same. Only the breadth of that awareness differs. We shall see that evidence from quantum theory led many of its founders and many current scientists to consider the possibility that the very existence of matter is dependent upon a conscious observer. Some of the evidence from near-death experiences discussed in chapter 6 of this book might be evidence that mind does *not* depend on matter for *its* existence.

Isn't it important therefore to consider the nature of mind as well as matter if we want to understand the relationship between mind and matter? The relationship between mind and matter is usually seen in one of two ways.

The first is the "materialist" view that mind is merely an "emergent" phenomenon that arises out of a certain arrangement of atoms that form brain cells in a particular juxtaposition and network. Within this view, sentience is merely an "onlooker" that has no role in determining our actions, and free will is an illusion.

The second is the dualist view that mind and matter are

fundamentally distinct and separate but that mind is able to influence matter through some as yet unexplained mechanism.

What if there were a third way? Could mind and matter be part of a continuum?

To define them in that way, we would need to define what mind and matter *are*. Many people might assume that of the two, the one that would prove the most difficult to define would be mind, as matter has presumably already been defined by science.

You may be surprised to know that it has *not* been defined!

As the theoretical quantum physicist Richard Feynman famously implied in his series of lectures "The Character of Physical Law," all that physics has done so far is to describe what matter does mathematically, but it hasn't revealed fundamentally what matter, space, and time actually are or why they interact as they do. This is perhaps analogous to someone who gets into an air traffic control tower and studies the positions and movements of the "blips" on the screen. That person could perhaps develop a timetable "law" governing the position, timing, and movement of the blips without ever being aware of the existence of airplanes or of pilots. However, I should say that following a recent experience of air travel, I do now believe that physical law is more reliable than airline timetables!

HUMANITY AND "ARTIFICIAL INTELLIGENCE"

I make the case in this book that artificial intelligence is an oxymoron. Intelligence implies understanding, and understanding implies awareness. I have shown in this chapter and in the others to follow that no machine can ever be aware. If you take a pocket calculator and type 1 + 1 =, then the display will show the number 2. However, your calculator has no awareness of the numbers 1 or 2. Instead, the answer arrives on the display as the result of the blind action of an algorithm. The calculator has no more conscious perception of addition than an abacus does. It is often said that Charles Babbage designed the first computer in the nineteenth century. However, in 1900, a team of sponge divers in Greece discovered the Antikythera shipwreck, and onboard this ship was a mechanical analogue computer that was thought to have been designed and made in Greece during the second or first century BCE. Some archaeologists believe that the computer was one of many treasures looted by the Romans that were on their way to be presented to Julius Caesar when the ship sank. It is now known as the Antikythera mechanism.

The British mathematician Alan Turing famously devised the "imitation game" in which a person interacts with a computer via a keyboard, with answers from the computer appearing on a screen. If the computer's responses suggest to the person that he or she is communicating with another person and not with a machine, then the computer, so Turing says, is displaying intelligence. If the computer were intelligent, then this of course would be artificial intelligence. As much as I respect Turing's intelligence, I dispute the notion that if we are convinced of something, then that means that it becomes true. If we find a horse with an artificial horn on its head and we believe we have found a unicorn, then would that mean that it is a unicorn or even that unicorns do exist?

Computers may indeed successfully imitate intelligence in the sense that we mistake them for sentient beings, but this mistaken belief does not imbue them with real conscious experience or even with consciousness at all.

It is true that we human beings may be easily duped into thinking that machines may be conscious, but that should not convince us that such a thing is even possible. Robots are being built that appear ever more humanlike in appearance, with soft skin and mechanical facial expressions. They may have "machine learning" such that they can adapt their behavior according to new information, but please bear in mind that the difference between human learning and machine learning is that humans are *aware* of what they learn whereas machines simply acquire more information. As I type these words on my computer, it retains them as information. It is not aware of that information, and it does not understand it.

One drastic and deadly way in which the belief in artificial intelligence could affect humanity is through the quest for mind uploading. In this book, I highlight the evidence from near-death experiences and other sources that shows that death is not an end for us as sentient beings. However, for those of a physicalist persuasion, we are thought to be just patterns of information in our brains, and so, they argue, if that information can be put into a computer, then our minds will exist in that computer. I know this may sound bizarre, and I believe it is bizarre also. If I kept a diary of all my experiences such that all my memories were written in a book, would that book become me?

The danger is that if people really believed this, then they might elect to end their life early and plan to continue in a machine or a robot that looked like a younger version of themselves.

Another danger has been highlighted by the Future of Humanity Institute at the University of Oxford and the Centre for the Study of Existential Risk at the University of Cambridge. This is that we may become the victims of the unanticipated consequences of what we instruct "artificially intelligent" computers to do. One example I have seen cited is that the computer could be tasked with executing a very complex mathematical problem that requires a huge amount of computing power. If it had machine learning, then it might deduce that it could increase its computing power by piggybacking on all the computers on the planet, but to do this it might deduce that the most efficient way would be to extinguish all human life on Earth (assuming that by that time the supply of energy did not rely on human intervention).[25] Other scenarios that are used to illustrate the point include one in which the computer is tasked with eliminating human suffering, which could also lead to our extinction as it would deduce that if there are no humans then there will be no human suffering.[26] None of this implies any malice on the part of the computer. It is incapable of malice or indeed of any other emotion. There are of course many other potential dangers, including the development of autonomous killing machines that could start off for use in warfare but would require only a small change in their programs (and remember machine learning could enable machines to alter their own programming) such that they might see all human beings as targets.

Many experts, including Professor Lord Martin Rees, the current British Astronomer Royal, have suggested that the organic or biological phase of a technological species such as ours may be a relatively short one, as once we start to develop artificial intelligence, it may be a very short time until it supplants and replaces us.[27]

Let me stress here that these experts are talking about nothing less here than the extinction of the human race. Perhaps if we can wake up soon enough to realize the limitless value of our humanity and why this cannot be replicated artificially, then there might be a chance for our species to survive.

This is the motivation for this book. Not a call to arms, but a call to hearts. Look around and see how you feel about those you love, and you will see that we must not allow you and them to become redundant and planet Earth to become mankind's tombstone.

The Universe and
the Mind's "I"

WHAT IS MATTER?

One would naturally expect any reasonable answer to this question to fit with those properties of matter that can be observed. Under certain circumstances, for example, matter can be visible and tangible. For example, this book is made of matter, and you can hold the book in your hands and see these words on the page. If you remember playing with magnets as a child, you will know that when two magnets are brought toward each other, you can feel a force—either pulling them together or pushing them apart. Simple experiments using only common household items can demonstrate the same properties related to electric charge or static electricity. Evidence was found early on in the twentieth century that demonstrated that atoms are almost entirely composed of empty space, as shown by the Geiger-Marsden experiment, for example.

At the center of the atom is the atomic nucleus, and at the edge of the atom are electrons. Both the nucleus and the electrons carry an electric charge. They each have opposite charges. By convention the nucleus is said to have a positive charge and the electrons to have a negative charge. This notion of positive and negative is a mathematical way of describing the fact that they are oppositely charged, so it wouldn't have mattered which way around they were defined as long as there is a consensus about which one is called positive and which one is called negative.

Magnets have north and south poles defined with reference to the alignment of the Earth's magnetic field. If you bring the like poles of two magnets together, you will feel a force pushing the two apart. However, if you bring the north pole of one magnet toward the south pole of another, you will feel a force pulling them together. Similarly, if you bring together two positively charged or two negatively charged objects, there will be a force pushing them apart. Again, if you bring together a positively charged object and a negatively charged object, there will be a force pulling the two together. Remember that the material of your hands and of the book is composed mainly of atoms that are themselves mainly empty space. Each atom has a positively charged center and a negatively charged edge. How does that make this book tangible in your hands? As your fingers press against the book, the outer edges of the atoms of both book and fingers are pushing each other away, and it is that resistant pressure that is perceived by you through the sense organs in your skin.

As you are reading these words, light is being reflected from the page and reaching the retina of your eyes. The light is interacting with some of the molecules in the retina, which through a chain of events is responsible for a signal being produced in the optic nerve, and that signal is transmitted to a part of the brain near the back of the head. Once the signal reaches this part of the brain, you are able to perceive the image of the words on the page consciously. The interaction of the light with your eyes is itself an interesting process. Light has been shown to be an electromagnetic wave, and its interaction with matter is related to the fact that matter also has electromagnetic properties.

You can also feel the weight of the book in your hands. The weight of the book is described in terms of the gravitational interaction between the mass of the atoms in the book and the mass of the Earth. This gravitational interaction between the book and the Earth pulls it toward the center of the Earth, but the electromagnetic force between the book and your hands or the table or whatever it is resting on prevents it from falling to the center of the Earth. We have seen that the matter that forms your body and the book mainly consists of atoms and that most of the volume of these atoms is empty space. At the center of each of these atoms is a tiny nucleus. The binding force that holds a nucleus together is described as the "strong nuclear force." You may have heard it said before that our bodies are made from "stardust." Our bodies and this book are composed of several different elements, the most prevalent of which in both cases are hydrogen, carbon, nitrogen, and oxygen. Elements are defined according to the composition of the nucleus of the atom. More specifically, they are defined by the number of protons in the nucleus (which is the same as the number of positive charges in the nucleus, as each proton carries a single unit of positive charge). Carbon, nitrogen, and oxygen have six, seven, and eight protons, respectively, in each atomic nucleus. The protons in the nucleus are held together by the strong nuclear force. The elements carbon, nitrogen, and oxygen are formed in stars as a result of nuclear fusion, and so it is true to say that all of these atoms in your body and in the book originated in stars. The formation of these atomic nuclei depends on the fact that they also contain chargeless or electrically neutral particles called neutrons. The formation of neutrons also happens in stars and occurs due to the "weak nuclear force." So your visual and tactile experience of this book derives ultimately from four types of material interaction: electromagnetism, gravity, the strong nuclear force, and the weak nuclear force. In fact, these four interactions are often described as the four fundamental forces of nature. However, it is important to realize that this description of the world in terms of matter or its interactions is just that—a description. Many people are surprised to hear that it does not explain what matter really is or why it behaves in this way!

The English astronomer, physicist, and mathematician Professor Sir Arthur Eddington had this to say about physicists' description of matter:

In physics we have outgrown archer and apple-pie definitions of the fundamental symbols. To a request to explain what an electron really is supposed to be we can only answer, "It is part of the A B C of physics."

The external world of physics has thus become a world of shadows. In removing our illusions we have removed the substance, for indeed we have seen that substance is one of the greatest of our illusions. Later perhaps we may inquire whether in our zeal to cut out all that is unreal we may not have used the knife too ruthlessly. Perhaps, indeed, reality is a child which cannot survive without its nurse illusion. But if so, that is of little concern to the scientist, who has good and sufficient reasons for pursuing his investigations in the world of shadows and is content to leave to the philosopher the determination of its exact status in regard to reality. *In the world of physics we watch a shadowgraph performance of the drama of familiar life.* The shadow of my elbow rests on the shadow table as the shadow ink flows over the shadow paper. It is all symbolic, and as a symbol the physicist leaves it. Then comes the alchemist Mind who transmutes the symbols. The sparsely spread nuclei of electric force become a tangible solid; their restless agitation becomes the warmth of summer; the octave of aethereal vibrations becomes a gorgeous rainbow. Nor does the alchemy stop here. In the transmuted world new significances arise which are scarcely to be traced in the world of symbols; so that it becomes a world of beauty and purpose—and, alas, suffering and evil.

The frank realisation that physical science is concerned with a world of shadows is one of the most significant of recent advances.[28] [my emphases]

Actually, we can say that all properties and interactions of matter can be described as force, but what is force?

The term *force* is used to describe a causative influence that results in change. In practice, in this context, we are referring to a net force. If you put the book on a table, it seems that there is a force due to gravity that provides the book with a weight by which it is "pressed down" onto the table. There is, though, an equal and opposite force exerted by the table on the book that supports the book's weight. As the two forces cancel each other out, the book can remain on the table, and its position (relative to the table) need not change. In this instance, the net force between the table and the book is zero.

We have seen that all interactions of matter can be described as force. The word *force* describes the causative influence by which matter is interconnected. It describes it but doesn't explain it.

Free will is the capacity to make an unforced decision between available options and is thus more likely to be found where force does not prevail. Force in its four manifestations (the strong and weak nuclear forces, electromagnetism, and gravity) can be seen as the driver, the instrument of cause and effect in a material scenario. Actually, a good case can be made that gravity should not be considered as a force, and I will elaborate on this in chapter 9.

Will can best be understood as primary, independent causation. If something is free, it has to have no force making it do what it does. So if we have free will, then this implies that we are able to cause something to happen without our choice being caused by anything.

So matter is characterized by being determined by force, whereas mind is characterized by having the capability to act independently of force through the exercise of free will.

Of the four fundamental forces described above, all except gravity have been described through the predictions of quantum theory. I find quantum theory fascinating when considered in the context of the ideas described in this book inasmuch as it implies, for many physicists, that consciousness or sentience has a crucial role to play in the physical universe.

Eugene Wigner wrote:

When the province of physical theory was extended to encompass microscopic phenomena through the invention of quantum mechanics, the concept of consciousness came to the fore again: it was not possible to formulate the laws of quantum mechanics in a fully consistent way without reference to the consciousness.[29]

Stanford University physics professor Andrei Linde wrote:

Will it not turn out, with the further development of science, that the study of the universe and the study of consciousness will be inseparably linked, and that ultimate progress in the one will be impossible without progress in the other? . . . Will the next important step be the development of a unified approach to our entire world, including the world of consciousness?[3]

In a recent video interview, Linde made a case that perhaps the reality of the physical universe itself is dependent on us as conscious observers.[30]

If this is correct, then could this perhaps suggest a clue to understanding the origin of the physical universe?

How did the universe arise from nothingness?

To quote Stephen Hawking:

There are something like ten million million million million million million million million million million million million million million (1 with eighty zeroes after it) particles in the region of the universe that we can observe. Where did they all come from? The answer is that, in quantum theory, particles can be created out of energy in the form of particle/antiparticle pairs. But that just raises the question of where the energy came from. The answer is that the total energy of the universe is exactly zero. The matter in the universe is made out of positive energy. However, the matter is all attracting itself by gravity. Two pieces of matter that are close to each other have less energy than the same two pieces a long way apart, because you have to expend energy to separate them against the gravitational force that is pulling them together. Thus in a sense, the gravitational field has negative energy. In the case of a universe that is approximately uniform in space, one can show that this negative gravitational energy exactly cancels the positive energy represented by the matter. So the total energy of the universe is zero.

Now twice zero is also zero. Thus the universe can double the amount of positive matter energy and also double the negative gravitational energy without violation of the conservation of energy. . . . "It is said that there's no such thing as a free lunch. But the Universe is the ultimate free lunch."[31]

(I hope those readers who have noticed that the word *million* is repeated once too often will forgive me as I have merely quoted the above paragraph verbatim from Hawking's book, which also contains this unimportant error!)

If the universe emerged from nothingness, then it also must have emerged from a state of existence in which there was no force. If there was no force, then what caused the universe to happen? If there was no force, no matter, no space, and no time, then what was there?

What if Schrödinger was right when he said that mind is indestructible and has no beginning or end?

If so, then mind would have been present in the initial no'thing'ness* from which the physical universe emerged. Could consciousness have played a role in the origin of the physical universe?

Hawking wrote an interesting comment about the relationship between theoretical physics and the physical universe.

Even if there is only one possible unified theory, it is just a set of rules and equations. What is it that breathes fire into the equations

---

*I have used the term *no'thing'ness* to denote the absence of "things" rather than the absence of existence. In the timeless nothingness from which physical universes emerge, there are no things, no substance, no space, time, or matter. Consciousness, however, is not a thing or a substance. In this book, I argue that its existence is independent of space, time, and matter and that consciousness is that which constitutes this primal nothingness. In the glossary at the end of this book I further clarify my use of terminology.

and makes a universe for them to describe? The usual approach of science of constructing a mathematical model cannot answer the questions of why there should be a universe for the model to describe. Why does the universe go to all the bother of existing?[31]

What if mind were the cause of the big bang? I should stress that this question need not rely on any notion of a creator god. Professor Freeman Dyson, who was based at the Institute for Advanced Study in Princeton, New Jersey, and was a colleague there to such luminaries as Kurt Gödel and Albert Einstein, wrote the following:

> It should not be surprising if it should turn out that the origin and destiny of the energy in the universe cannot be completely understood in isolation from the phenomena of life and consciousness. . . . It is conceivable . . . that life may have a larger role to play than we have imagined. Life may have succeeded against all odds in molding the universe to its purposes. And the design of the inanimate universe may not be as detached from the potentialities of life and intelligence as scientists of the twentieth century have tended to suppose.[32]

The notion of the big bang origin of the physical universe implies that the universe came from a state of nothingness via an initial singularity. This was the origin of space, time, and matter. When comparing the universe as we know it now to the initial singularity, one of the fundamental differences can be summed up as the contrast between union and separation.

Whereas a state of union could be independent of space, time, and matter, a state of separation would seem to be virtually synonymous with them. What is space but the separation of points? Some people might challenge the notion that the nothingness from which the physical universe emerged should be described as a "union." "Surely," they might say, "to have a state of union, there must be things or parts together for these to be united. If nothingness has no contents, then what is united?" My answer is that in the timeless state that I have referred to as no'thing'ness, all sentience is one. We have seen that Schrödinger made a case that time is a product of consciousness and therefore that sentience cannot begin or end in time (see page 40). As sentient beings, we would therefore be eternal and not begin at conception or birth and also not be ended by physical death. Of course, the obvious question that arises from this would be "where" were we before and "where" will we be afterward. Perhaps part of the answer lies in the notion of reincarnation.

I have argued that we have taken on physical bodies as a result of becoming limited and therefore separated from each other by our individual and unique patterns of restriction. If that were so, then it would follow that if the limitation remains, then we might return in "another life" after we have died because the process of taking on another physical body could repeat itself if we have not resolved the problems that made it happen on the previous occasion. Our identities are determined by what we have made of ourselves.

The more we divide and act out of "selfishness," the more we limit ourselves to one point and therefore enhance our restriction and distance from our full potential as "the all." Whenever I hear people talk about the nature versus nurture debate, I always consider what I call the "nature *and* nurture plus neither" debate! What if one of the factors that influences who and what we are is what we have made of ourselves through what we have been and done *before* we were born or conceived in this particular lifetime? Similarly, the way that we live *now* might influence the manner in which we continue after death, which could possibly include any future lives we might have. When I talk to people about this, they often ask why, if that is the case, we do not remember previous lives, and I also have often heard people assert that if we do not remember those lives, then it is not meaningful to claim in any way that those previous lives were really "ours." I must confess to finding such an assertion to be erroneous and lacking in substance. If tomorrow you developed amnesia, would that mean that today you are not you? If you regained your memory, would you then become you again? By "you," I do not mean your name or your perception of your identity, but rather I mean your continuous "stream of consciousness."

Research in sleep laboratories suggests that when we sleep, we dream several times during the course of that sleep, whenever we reach the rapid-eye-movement (R.E.M.) phase of the sleep cycle. However, we only tend to remember the dream if we awake during that particular instance of R.E.M. sleep. If you are only defined by conscious memories, then wouldn't that mean that it was not really you who had the dreams that you forgot? If it was not you, then who was it? There are many documented examples of people who do have recollections of previous lives that include memories of events or facts that they could not have known about as part of their experience in this life but that are later verified after they have gone on the record about their memories.[33] It is interesting that in many of these cases, the life that they recall had a sudden and often violent end. Bearing in mind that we tend only to remember those dreams from which we are awoken, I often wonder whether a similar mechanism might be relevant here.

I must stress though that in comparing a physical life to a dream, I am in no way suggesting that our human existence is in any way "virtual" or "imagined." Actually, such suggestions are anathema to me and remind me of the musings of "armchair" philosophers. If they get up from their chairs and stub their toes on the way to the toilet, then one would hope that they are reminded that their dream is all too real! The danger of solipsism is that it has a lot in common with psychopathy. If others are merely part of your dream, then why have a conscience?

Incidentally, you may hear people say that reincarnation could not be true as the number of people alive in the present day is greater than the number of all the people who have died in the past. Setting aside for a moment the *logical* flaws in that argument, I would like to point out that it is also factually misleading. Joel Cohen of the Population Reference Bureau (a nongovernmental organization based in Washington, D.C.) estimated that the total number of people who have ever been born is around 106 billion![34]

Evidence from near-death experiences (see chapter 6) provides further support for the notion that we do not end when we die. If our sentience did not begin but the universe did begin, then that implies that as well as having existence within the physical universe we must also have a timeless, eternal existence in the state beyond space, time, and matter from which the universe emerged. In that state of existence, there is no space or time and therefore no separation, and it would therefore follow from this that we must be "one" in that state. Seen in these terms, perhaps the suggestion that we "love our neighbors as ourselves" is a clue to our origins within a state of union and wholeness where there is no pain and no decay and no matter. It would then also be an indication of how we might be able to return to that state and perhaps rejoin the light that numerous people see in their near-death experiences (again, see chapter 6) that they describe as embodying wisdom and compassion.

So, to summarize, the origin of the physical universe from a state of no'thing'ness could not have been instigated by force, as force only exists in the context of space, time, and matter, which themselves were brought into being by the big bang. I have presented a case that free will is the one causative influence that does not require the preexistence of a physical universe. We have seen that the physical universe is characterized by the separation of points and therefore by space and time but that the singularity from which it emerged represents a pole of initial union from which this began.

Putting these ideas together, one could speculate whether perhaps the physical universe itself could have begun as the result of a choice of sentient beings to experience separation and therefore to become separate (from each other). Matter, space, and time are simply the products and properties of this separation. I know this is a big claim, but there is big evidence to back it up in modern physics! Linde was so impressed by evidence pointing to a central role of consciousness in the origin of the universe that he included mention of it in his much-acclaimed textbook on cosmic inflation. He has suggested also that it seems possible that mind could be the creator of matter (see quote on page 62).[35]

Remember that Schrödinger also suggested that:

I—I in the widest meaning of the word, that is to say, every conscious mind that has ever said or felt "I"—am the person, if any, who controls the "motion of the atoms" according to the Laws of Nature.[2]

My suggestion is that mind and matter form part of a continuum. I have made a case that the power of the will is none other than that of primary, independent causation and is thus the power of all power. If "we" have chosen to exist in separation, then people might ask, What happened to the state of union from which we came? Has it been broken apart or ceased to exist? Remember, time begins with the physical universe, and time has no meaning without separation. The implication of this is that the prior state of all knowing and all potential is not temporal and therefore does not change. This would imply that in some very real sense, we currently "coexist" within two states at once. We are simultaneously the "one" and the "many." This may seem counterintuitive, that something can be in two diametrically opposed states at once, but if so, then please spare a thought for Schrödinger's poor cat!*

Whereas free will implies primary causation, physical forces, such as magnetism and gravity are properties of secondary causation. By this I mean that force itself also has a determined cause in accordance with natural law. *Force* describes the interactions of matter. *Will* describes the propulsion of mind or sentient being.

How might the two be related?

There is a sense in which the two might be seen as opposites. This is because, by definition, if a choice is made freely, then there can be no

_____

*If you're unfamiliar with Schrödinger's thought experiment, see page 137.

force that impels it to be such as it is. So it would seem that free will, if it exists, would be most likely to be found where there is no force. The nothingness, or as I would call it, the no'thing'ness, from which the physical universe emerged would of course be the clearest example of this.

Schrödinger also wrote about the unitary nature of consciousness.

> Knowledge, feeling, and choice are essentially eternal and unchangeable and numerically one in all men, nay in all sensitive beings. But not in this sense—that you are a part, a piece, of an eternal, infinite being, an aspect or modification of it. . . . For we should then have the same baffling question: which part, which aspect are you? What, objectively, differentiates it from the others? No, but, inconceivable as it seems to ordinary reason, you—and all other conscious beings as such—are all in all. Hence, this life of yours . . . is, in a certain sense, the whole.[2]

It's interesting to contrast this unitary nature of consciousness with the nature of physical reality, which would *seem* to be fundamentally divided, for example, into separate particles of matter, locations in space, and moments of time. Also, mind seems to be divided into units that we know as individuality—"you" and "me" and every"one" else. If our individuality is a derivation from an underlying unity, then how about the separation of locations of space, time, and matter that we perceive as physicality?

··

Schrödinger's Lion

Empirical experiments have proven that quantum mechanics shows that matter cannot have both of two properties that are known as "reality" and "separability."[36] Used in this context, the term *reality* refers to the existence of characteristics that would be present whether or not they were observed or measured. These include properties such as position and momentum. Experiments in quantum mechanics have shown, for example, that electrons don't have a location until they are observed. It is because of this that they can simultaneously have wave and particle characteristics and can behave as though they are in several places at once. This property of simultaneously being two different seemingly contradictory things is of course what Schrödinger was alluding to in his thought experiment with the cat (see page 137). Also, the phenomenon of quantum entanglement can be demonstrated in such a way that shows that two particles separated in space can show properties that prove that at a fundamental level they are one.

In 1982, a remarkable experiment was carried out by Alain Aspect and his team at the Institut d'Optique Théorique et Appliqué in Paris. What they found was that when a pair of "correlated" photons (particles or units of light) were sent in opposite directions, a measurement of one of the photons could alter the state of the other photon, even though there was insufficient time for a signal to travel from one photon to the other.[36,37] This was as predicted by the equations of quantum mechanics, but it also was another experimental confirmation that these equations were correct and provided empirical evidence that the properties of the photons do not exist until they are measured and also that dividing the photons in space and time does not really separate them. Thus are falsified the notions of objective "reality" and of the separability of the photons. Technically, the experiment implies merely that reality and separability cannot coexist, so we must dispense with at least one of these notions. Later experiments appeared to show that we must dispense with both the notion of a (observer independent) reality *and* the notion of separability.[38]

As I have used terms such as *quantum entanglement,* I would like to mention that I take issue with those who suggest that conscious awareness is the product of quantum entanglement. After all, the act of observation in quantum physics is that which collapses the wave equation and therefore destroys entanglement! It's a bold claim, I know, but I would maintain that in this book I demonstrate that conscious awareness is not the product of any "thing" but is indeed the ground of existence itself from which all "things" begin and end. This being so, it is apparent that it could never be possible either to manufacture a mind or to upload one into a computer.

....

The unitary evolution of the Schrödinger wave equation implies that one can consider the whole physical universe as a unified quantum state when applying the *"wave function of the universe."*[3]

The "wave function of the universe" is a subject that has been considered by such notable cosmologists as Stephen Hawking and Andrei Linde.[3,30,31,35] As Linde has pointed out, when the wave function of the entire universe is considered, one realizes that the presence of conscious observers is required to "bring the Universe to life."[3,30,35]

Therefore if one would wish to describe the evolution of the universe with the help of its wave function, one would be in trouble: The universe does not change in time, it is immortal, and it is dead.

The resolution of this paradox is rather instructive. The notion of evolution is not applicable to the universe as a whole since there is no external observer with respect to the universe, and there is no external clock as well which would not belong to the universe. However, we do not actually ask why the universe as a whole is evolving in the way we see it. We are just trying to understand our own experimental data. Thus, a more precisely formulated question is why do we see the universe evolving in time in a given way. In order to answer this question one should first divide the universe into two main pieces: an observer with his clock and other measuring devices and the rest of the universe. Then it can be shown that the wave function of the rest of the universe does depend on the state of the clock of the observer, i.e., on his "time." This time dependence in some sense is "objective," which means that the results obtained by different (macroscopic) observers living in the same quantum state of the universe and using sufficiently good (macroscopic) measuring apparatus agree with each other.

Thus we see that by an investigation of the wave function of the universe as a whole one sometimes gets information which has no direct relevance to the observational data, e.g., that the universe does not evolve in time. In order to describe the universe as we see it one should divide the universe into several macroscopic pieces and calculate a conditional probability to observe it in a given state under an obvious condition that the observer and his measuring apparatus do exist. Without introducing an observer, we have a dead universe, which does not evolve in time. Does this mean that an observer is simultaneously a creator?[3,30]

So in the above quotation, you can see that one of the world's leading physicists points out that after considering the evidence, it seems that you, the reader, as a sentient observer are all it takes to make the whole physical universe exist. Many people think that science tells us that we are insignificant lumps of matter that have randomly appeared and that when our bodies die we just cease to exist. And yet here we see that the evidence actually suggests the opposite. Sentient beings like you and those around you are actually the most significant and powerful "things" in the whole universe.

We have seen that Schrödinger made a case that at a fundamental level all consciousness has a single root, and we have also seen that his equation, when applied to the whole physical universe, implies that it is only the separation between subject and object, that is, between observer and observed, that makes it possible for time and physical reality to exist.[2] My suggestion is simply that the separation between subject and object that generates physical reality is fundamentally the same as the separation between subject and subject that creates "separated individuality." In the chapters that follow, I will explore this possibility in more depth and consider the implications it might have for how you and I exist as sentient beings and what our true potential might be.

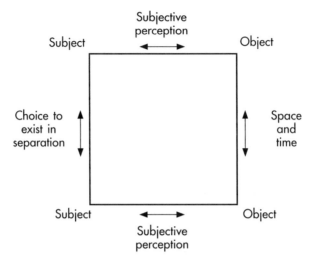

Figure 4.1. The figure above is intended schematically to depict the notion that object and object are separated (by space and time), that subject and object are separated (by subjective perception), and that subject and subject are separated (by "self-centeredness" or the "choice to separate").

WHAT IS MIND?

In the foregoing, I have addressed the question of the nature of matter and demonstrated some of the evidence that shows that matter depends on mind for its existence. Without an observer, there is no reality, no time, no space, and no matter. There are only possibilities and no "actuality."

What constitutes an *observer*? An observer can only be an observer because he or she is sentient. The observer has *awareness*. So what is awareness?

When people define something, they always do so in terms of *other* things, and so the definition is *relative*. One may define houses in terms of bricks, bricks in terms of clay, clay in terms of chemistry, chemistry in terms of physics, physics in terms of mathematics, and so on.

But what makes a house a home? A house is a home because of what it means to those who *live* there and also what those people *mean* to each other.

Usually, when we define something, we do so by reducing it to its constituent parts and defining these parts in relation to other parts.

What parts can we reduce awareness to? Well, we could reduce it to our*selves,* to our individuality. We could reduce mind to *minds,* but where does this take us? We are still left with the problem of how to define our individual subjective awareness. I must repeat here that I am not referring to the *content* of our awareness, not describing what we are aware *of,* but rather I am referring to the fact that we are aware *at all.*

Many people say that this merely relates to the fact that we can consider the observation that we are perceiving. They would argue that somehow mind is a product of self-reference, which generates self-awareness, and that this is a prerequisite for awareness of anything else. However, there is a danger here of falling into a trap of specious argument.

As I have pointed out earlier, if I point a video camera at a mirror and press the record button, then it will process information about itself and the fact that it is recording information about itself, but this does not make it *aware!*

The relative definition of awareness works something like this: I am human. I know that I have a mind. I see that you are human, and therefore I understand that you also have a mind, that there is someone whom I call "you" that is perceiving the world from "behind your eyes." In the final analysis, we understand awareness because we are aware, and we cannot define that awareness to others except by referring them to their own awareness. You can explain this to another person but not to a machine. As it has no awareness, it cannot understand what awareness *is.* (Actually, a machine or a computer cannot understand anything at all, as understanding requires awareness.)

This is not a cop out. I am not trying to escape a definition of what mind is, but rather I am addressing the elephant in Searle's Chinese room.

Awareness is the fundamental ground of all existence and can never be reduced to any 'thing' else!

If awareness is fundamental, then why does it seem to be expressed or perceived separately by *individuals* such as you and me?

When we look around us, we don't see unity. We are all aware of the separate individuality of ourselves and others. We are aware of apparently separate physical objects and separation between locations in space and locations in time: "here" and "there," "then" and "now."

Why does this separation exist?

The Origin of Separation

*What Are Space, Time, and Matter, and
Why Do They Exist?*

We are used to conceiving of time as "flowing," and many people use analogies such as that of a flowing river, the implication being that each location on the path of the river is likened to a moment in time. However, it is relatively easy to see that this metaphor breaks down as soon as we realize that for time to flow, it would have to have a medium in which to flow as well as a speed or rate of flow. Does time move at one second per second? To say that time flows at one second per second would be as silly as saying that space flows at one meter per meter. It is meaningless.

It is interesting to consider two aspects of time that need to be accounted for by any theory that claims to explain what time is. These are the existence of the *present,* as in the past, present, and future, and the "arrow of time."

The *arrow of time* is a phrase used to denote the distinction between past and future in the sense of the direction of change as we experience it. This also encompasses the notion of causality in the sense of causes "preceding" their effects.

Schrödinger observed, as we have seen, that "mind is always *now.*" I would suggest that we could elaborate on this by saying that "now" only exists for sentient observers and also that it is the very act of sentient observation that creates the "now"!

Seeing the Past

A clear distinction needs to be made between the *content* of "now" and the fact of our conscious perception of it within the present. To take one example, when you look out at the night sky, you see stars and they form part of your present experience at that moment. However, the light by which you are seeing those stars began its journey many, many millions of years ago.

Let's consider the path of the ray of light that came from the star to your eye. It is particularly interesting to consider this within the context of Einstein's theories of relativity. You may be familiar with the notion that light always moves in straight lines. Well, according to Einstein's theory (and this is backed up by empirical observation), space itself is curved. So as light travels straight through curved space, it can be inferred to be "bent" by gravitational fields. In certain contexts, this is known as "gravitational lensing," which has actually been observed. Where there is a distant galaxy that has a large cluster of galaxies in between it and us, then we can see that there is a magnification and distortion of the image of the background galaxy that we see through our telescopes as though the galaxy cluster is acting like a lens, which indeed it is. In fact, in 1919, Sir Arthur Eddington achieved public recognition during an observation of a solar eclipse in which he demonstrated that the sun acted as a gravitational lens, thereby confirming Einstein's theory of general relativity. The lens effect can be shown to be due to the effect of gravity as a distortion of space-time.

There is another fascinating property of light and its relationship to space-time that was discovered by Albert Einstein. The story goes that the young Einstein, aged just sixteen, conducted one of the first of his "thought experiments," or *gedankenexperiments*. He imagined a scenario where he was traveling at the speed of light while holding a mirror, which was of course also traveling at the speed of light. Prior to discovering his theory of relativity, he wondered whether his image in the mirror would disappear. If space and time had been absolute and not relative, then you could imagine that the light from his face would not have a chance to reach the mirror, as to do so it would need to move faster than the speed of light to get there. He knew that the speed of light was a constant from physicist James Clerk Maxwell's earlier work on electromagnetism. Einstein was once asked whether he had metaphorically been standing on Isaac Newton's shoulders when he discovered relativity, and he replied "No, I stand on Maxwell's shoulders!"

....

Einstein's great discovery was based on the realization that the speed of light is the same for all observers. Speed, though, is a measure of distance traveled per unit of time, and a consequence of relativity is that space and time are not merely an inert background against which things happen and move but that they are themselves related to what is happening. For example, as you get closer to the speed of light, space and time become compressed. An example that is often cited is the "twin paradox." The twin who accelerates up to almost the speed of light experiences less passage of time during the round trip to a distant location than the twin who stays on Earth, such that when they reunite, the traveler has aged far less than the one who stayed at home. It turns out that as you approach the speed of light, distances in space and intervals of time shorten. Imagine you were wearing a watch while accelerating to a speed close to the speed of light. To you, the watch would appear to be ticking at its normal rate, but if you were able to see your twin back on Earth, his or her clock would appear to be whizzing through the minutes and hours far more rapidly than you would expect. If your twin could see your watch, then he or she would see it seeming to be ticking very *slowly*. Now, material objects with mass can never reach the speed of light, so we can never actually accelerate a tape measure or a clock to the speed of light, but if you imagine a hypothetical clock or tape measure associated with a light beam, then Einstein's equations suggest that the path of the light beam as viewed from within the reference frame of the light beam has no length and no passage of time. In other words, from the perspective of light, there is no separation between points in space or "moments" of time.

Take the example of light that left a distant galaxy billions of years ago and that reaches your eyes now. From the perspective of the light beam, those billions of years are all instantaneous, and the billions of light years between you and the distant galaxy are null. So there is no separation between the origin and the destination of the light beam in space or in time. Spatial and temporal separations are nullified from the perspective of light. In fact, diagrams of Minkowskian space-time (which is a mathematical description of Einstein's special relativity) depict light traveling along paths known as "null cones." The word *null* in this context refers to the fact that distances in space and intervals of time are zero from within the reference frame of light on the null cone.

In January 1991, when I was at medical school, I attended a public lecture given at the university in Scotland where I was studying by the eminent physicist Professor Sir Hermann Bondi on the subject of Einstein's theory of relativity. There was an opportunity at the end to put questions to Bondi. I was very happy to have such an opportunity, and I asked him about Einstein's time dilation effect. I asked him whether I was correct in surmising that Einstein's equations suggest that if one could travel at the speed of light, then one could be in a frame of reference in which subjectively no time passes and one would theoretically be able to be "everywhere at once." He replied that the equations also showed that mass increases as one approaches the speed of light. As mass equates to the resistance to further acceleration and mass approaches infinity as one approaches the speed of light, he said that it is impossible for an observer to reach the speed of light because observers have mass.

Einstein's theory of relativity does indeed imply that to move at the speed of light requires that that which is moving (e.g., a photon) has to have no mass, or more specifically, zero "rest mass." Massless particles move at the speed of light, and therefore no time passes within their reference frame. Massive particles move at speeds slower than that of light, and what we call time derives from this. My contention was that to make definitive statements about observers or sentience, we would need to have a clear understanding of the nature of mind and what its physical correlates might be. Although we are used to considering minds as associated with physical bodies that have mass, such as human beings, I did not accept that it is an established fact that mind could only exist in that context. As we will see in chapter 6, empirical evidence from scientific studies of near-death experiences shows that mind need not necessarily be confined to being associated with a physical body. In fact, quantum theory demonstrates, according to some quantum physicists, that matter is made real through consciousness as a result of the process of observation. If matter is made real only through the mind, then how can the mind be seen as just a property of an arrangement of matter?

My reply to Bondi was to ask, If we do not know the mass of "thought," then why is it necessary to assume that minds could only exist in association with bodies that have mass? If it were possible for an observer to be like light and have no mass, would my original question be answerable in the affirmative and would it be possible for a sentient being to be everywhere at once?

He replied that it would be possible.

Bondi had originally moved to Cambridge from Austria, as he had wanted to work with Eddington, who was the Plumian Professor of Astronomy at the University of Cambridge. In a letter of recommendation to the university, Eddington described Bondi as "a mathematical student of great brilliance and promise."

There is an interesting anecdote about Eddington, who was a contemporary of Einstein and was considered an authority on relativity. During one of Eddington's lectures, he is said to have been told by the Polish-American physicist Ludwik Silberstein, "Professor Eddington, you must be one of three persons in the world who understands general relativity." Eddington paused, unable to answer. Silberstein continued, "Don't be modest, Eddington!" Finally, Eddington replied, "On the contrary, I'm trying to think who the third person is."[39]

One reason why I mention Eddington is that he had interesting views on the nature of mind and the relationship between consciousness and the physical universe. In 1928, he published his book *The Nature of the Physical World*, which was based on the Gifford Lectures that he had delivered at the University of Edinburgh in 1927. There is a chapter on the "new quantum theory" in which is to be found the following gem:

> The universe is of the nature of a thought or sensation in a universal Mind. . . . To put the conclusion crudely—the stuff of the world is mind-stuff. As is often the way with crude statements, I shall have to explain that by "mind" I do not exactly mean mind and by "stuff" I do not at all mean stuff. Still that is about as near as we can get to the idea in a simple phrase. The mind-stuff of the world is something more general than our individual conscious minds; but we may think of its nature as not altogether foreign to feelings in our consciousness. . . . Having granted this, the mental activity of the part of the world constituting ourselves occasions no great surprise; it is known to us by direct self-knowledge, and we do not explain it away as something other than we know it to be—or rather, it knows itself to be.[28]

So the separation of points as seen in terms of space, time, and matter is only relevant where there is mass. From the perspective of a massless "light beam," source and destination are not separated in space or in time. Soon after the big bang was a period known to cosmologists as the photon-dominated era. After this, more energy became condensed into the form of mass. According to Roger Penrose, the final stages of the universe will also be photon dominated, and eventually mass will disappear and there will be no more "passage of time" until perhaps a new big bang will occur from out of the remnants of this physical universe to form another physical universe.[40]

Einstein's equation $E = mc^2$ implies that mass is a condensed form of energy (see, for example, the article titled "Antimatter" on the CERN website[41]), and it is known that under certain circumstances matter can be generated from the energy contained within electromagnetic radiation. Also, the converse can occur, as, for example, when an electron meets its antimatter equivalent (a positron) and both are annihilated as the energy is released, giving rise to electromagnetic radiation.

The American theoretical physicist David Bohm made an interesting observation about light during a conversation with professor emeritus of Rutgers University Renée Weber. He was asked "Do you have any hypothesis as to why light has been singled out as the privileged metaphor [in mysticism]?" He replied as follows:

> If you want to relate it to modern physics (light and more generally anything moving at the speed of light, which is called the null-velocity meaning null distance), the connection might be as follows. As an object approaches the speed of light, according to [the theory of] relativity, its internal space and time change so that the clocks slow down relative to other speeds, and the distance is shortened. You would find that the two ends of the light ray would have no time between them and no distance, so they would represent immediate contact. (This was pointed out by G. N. Lewis, a

physical chemist, in the 1920s.) You could also say that from the point of view of present field theory, the fundamental fields are those of very high energy in which mass can be neglected, which would be essentially moving at the speed of light. Mass is a phenomenon of connecting light rays which go back and forth, sort of freezing them into a pattern.

So matter, as it were, is condensed or frozen light. Light is not merely electromagnetic waves but in a sense other waves that go at that speed. Therefore all matter is a condensation of light into patterns moving back and forth at average speeds which are less than the speed of light. Even Einstein had some hint of that idea. You could say that when we come to light we are coming to the fundamental activity in which existence has its ground, or at least coming close to it.[42]

It is interesting that people like Siddhartha Gautama (Buddha) and Jesus of Nazareth are often depicted as having had light around them and that there are reports that at times Jesus's body was seen to shine, for example, at what is referred to as the "transfiguration." I have suggested that perhaps it was the initial choice to exist in separation that froze light into matter around the time of the big bang origin of the physical universe. If that is so, then is it conceivable that the opposite of this achieved through what Jesus referred to as loving the neighbor as self actually begins to unfreeze matter, which then begins to shine? If so, could the light occurring as a result of this be of the same nature and origins as the light that left his image on the Turin Shroud?

I should stress that I take caution not to attribute too much strength to anecdotal reports about individuals in history, but also that the image on the Shroud is an empirically verified phenomenon and that there is evidence pointing to a burst of radiant energy from the dead body that was within the Shroud having been the cause of the image.

THE ARROW OF TIME

The *arrow of time* is a phrase we use to denote the apparent direction of time, with causes preceding their effects and the direction of change being toward increasing disorder. Eddington actually coined the phrase in his 1928 book *The Nature of the Physical World* when he wrote:

> Let us draw an arrow arbitrarily. If as we follow the arrow we find more and more of the random element in the state of the world, then the arrow is pointing towards the future; if the random element decreases the arrow points towards the past. That is the only distinction known to physics. This follows at once if our fundamental contention is admitted that the introduction of randomness is the only thing which cannot be undone. I shall use the phrase "time's arrow" to express this one-way property of time which has no analogue in space.[28]

It is often suggested that the arrow of time arises from the second law of thermodynamics, which states that in a closed system, disorder will increase. However, this argument is derived from statistical mechanics, and since the events this describes are all reversible, it can be used to suggest that we should expect to see disorder increasing in the "future" direction but also in the "past" direction. This is a counterintuitive notion but can be confirmed easily using basic mathematics. Strange as it may sound, if not for the fact that we know that order proceeds from order, we would find it to be more likely that it appears as a random fluctuation. The whole glass on the table before it falls to the floor is, in this sense, more likely to have appeared as a result of broken fragments of glass gathering together and rising up to fit precisely together than the glass is to have been manufactured! In other words, the fact that we observe a continual increase in disorder, or *entropy*, with time is simply a pointer to the incredibly ordered origin of the physical universe.[40]

Perhaps it would be worthwhile to define *order* in the context of this discussion. Actually, *order* is only defined by implication in physics, as entropy is a measure of *disorder*. *Order* could therefore be defined as a low-entropy state.

A zero-entropy state would refer to a state with only one possible configuration, which could be a state of nothingness as there are no "things" to configure!

....

The "Arrow of Time"

Imagine if you had a glass on your table and it fell to the floor, shattering into many pieces (see fig. 5.1). There are few ways of having all those pieces of glass together as one coherent piece of glass but many ways for them to be separate or broken. For this reason, it is far more likely for a glass to break into many pieces than it is for many separate shards and crumbs of glass to assemble themselves

Figure 5.1. An intact glass and the same glass shattered. If not for the evidence that the universe began from a state of ultimate order, it would be statistically far more likely for a glass to appear spontaneously out of shattered shards than for it to be manufactured. Image courtesy of Daniel Langsman.

spontaneously into one whole glass. In fact, the latter is so unlikely that you would almost certainly need to wait longer than the lifetime of the universe for one instance of it to happen. However, the physical laws that govern the movement and interactions of the silicon and oxygen atoms that the glass is made of are reversible in time, so, being faced with the broken glass and "playing the film backward" to see the pieces assemble themselves together into one glass should also be just as surprising unless we presuppose that in the direction of time that we generally call "backward," all of space, time, and matter would be seen to move toward an end point of an extremely high degree of order. Although it is so bizarrely improbable that a broken glass would spontaneously assemble itself into a whole one, it is still far more probable than the random appearance of a universe that began with sufficient order that would enable the emergence of the whole glass several billion years later.

Actually, it has been calculated that for the observable universe to have begun in as ordered an initial state as it did, the odds are so immense that it is difficult to conceive of the numbers involved. For example, in betting parlance, people talk about odds of two to one or one hundred to one and so on. As you start to add extra zeroes on the end, the odds rapidly escalate so that odds of a billion to one are much higher than odds of ten to one, even though you have only added eight zeroes. Well, to imagine the odds for the universe to have begun in as ordered a state as it did, you could try adding one zero for each and every atom in the universe and you would still not even come close to a number big enough to express the odds![23]

However, if we suggest that the universe began from an initial pre-existent state of union in what I have called no'thing'ness, then this would be consistent with a zero entropy "boundary condition" for the physical universe.

Some cosmologists have proposed that the apparent immensely ordered origin of the observable universe arose from a process called *inflation,* which smoothed out the unevenness and irregularities from our part of the universe at a very early stage, soon after the big bang. However, others, such as Penrose, have pointed out that this is fudging the issue, as one has to assume an even higher degree of fine-tuning and order for inflation to be able to achieve this.[23]

If, as I suggest, the physical universe began as a result of a choice to separate, then it is perhaps this choice to separate that demarcates the arrow of time as long as sentient beings remain in separation within this physical universe. With this in mind, it is interesting that the man of the Shroud (if we are to accept the plentiful evidence suggesting that the man whom the Shroud once wrapped was none other than Jesus of Nazareth) is reported to have made a statement that in retrospect would appear to refer to entropy and the arrow of time. This is reported in Matthew 6:19–21.

> *Lay not up for yourselves treasures upon earth,*
> *where moth and rust doth corrupt, and where thieves*
> *break through and steal:*
> *But lay up for yourselves treasures in heaven,*
> *where neither moth nor rust doth corrupt, and where*
> *thieves do not break through nor steal:*
> *For where your treasure is, there will your heart*
> *be also.*

He seems to have suggested that to be physical and temporal is to exist in a state where everything decays, but that there is another frame of existence that we could potentially access and achieve where time's arrow and entropy do not prevail. Again, I think it bears repeating that these words may be open to rational verification, which might lead us to a revolutionary way of understanding ourselves, each other, and the world we live in. It's a shame that his words appear to many people to have been hijacked by religious institutions. My claim throughout this book is that we don't need to rely on faith or "mystery" or institutional "authorities" to understand what he said and did, but instead we all have an equal right to consider his words and actions through reason and evidence. The Shroud of Turin might well be evidence that he did indeed practice what he taught, and the very next verse, in Matthew 6:22, might be a clue to all of this:

The light of the body is the eye: if therefore thine
eye be single, thy whole body shall be full of light.

Remember that according to Einstein's relativity (see the quote from Bohm on pages 73–74), light would appear to be that for which the separations of time and space are resolved, as no time or space passes from the perspective of being light. How might this relate to the puzzle of our perception of time?

The following are my suggestions for the origin of the "now" and the arrow of time:

The present, or the "now," is nothing more and nothing less than our "presence" as sentient beings. We are always *now*, and all our experience exists within the present (including memories of the past and expectations about the future).

The "arrow of time" is simply the distinction between the past and the future. The future is not fixed because free choices are never predetermined!

In his book *The Trouble with Physics,* physics professor Lee Smolin of the Perimeter Institute makes the following observation about time while discussing how to unite relativity with quantum theory:

I believe there is something basic we are all missing, some wrong assumption we are all making. If this is so, then we need to isolate the wrong assumption and replace it with a new idea.

What could that wrong assumption be? My guess is that it involves two things: the foundations of quantum mechanics and the nature of time. . . . But I strongly suspect that the key is time. More and more I have the feeling that quantum theory and general relativity are both deeply wrong about the nature of time. It is not enough to combine them. There is a deeper problem, perhaps going back to the origin of physics.

Around the beginning of the seventeenth century, Descartes and Galileo both made a most wonderful discovery: you could draw a graph, with one axis being space and the other being time. A motion through space then becomes a curve on the graph. In this way, time is represented as if it were another dimension of space. Motion is frozen, and a whole history of constant motion and change is presented to us as something static and unchanging. If I had to guess (and guessing is what I do for a living), this is the scene of the crime.

We have to find a way to unfreeze time—to represent time without turning it into space. I have no idea how to do this. I can't conceive of a mathematics that doesn't represent a world as if it were frozen in eternity. It's terribly hard to represent time, and that's why there's a good chance that this representation is the missing piece.[43]

Perhaps the problem is that we are trying to represent time in an explicit way as existing outside the sentient observer. However, this is a contradiction in terms because time (both in the sense of the "now" and the arrow of time) is only made real by and within the sentient observer!

If you look around you now, you might see what appears to be a substantial world. A world of things. Matter with extension in space existing "now" and moving or changing with the passage of "time." The behavior of matter is generally very accurately predicted by quantum mechanics. However, Schrödinger's wave equation describes a linear, fully deterministic evolution of the quantum state that implies that the properties it describes, such as position, momentum, and so on, do not have defined values, so it would be meaningless to talk about the electron being "here" or "there," for example. Somehow, consciousness seems to be implicated as the agency that makes the wave equation "collapse," which thereby creates temporal reality. The accuracy of quantum mechanics is actually not in describing how the world is but in telling us what we are likely to see if we look at it! This may sound like a nonsensical distinction, as you might assume that it would be the same whether or not we look at it. You would be in good company if you think that, because Einstein himself had the same objection. However, in light of empirical evidence, he lost the argument with Niels Bohr on that. Simple tabletop experiments can confirm that when an electron or a photon (or, in some experiments, molecules made up of dozens of atoms called "buckyballs") is not observed, it will behave as though it may have been in more than one place at a time, so that it will form patterns that show that one particle was acting as a spread-out wave, but this wave is not a substance or a disturbance within a substance but instead relates to the probability of where you would have found it if you had looked for it while it was traveling!

Light or anything else that has no mass appears from our vantage point to move at the speed of light, and in so doing, from its own frame of reference, joins together points that usually appear to us, from within our perspective in physical life, to be separated in time and space.

If Bohm was right and matter is frozen light, then what are the implications for our existence as physical human beings, as amalgams of mind and matter?

The origin of matter would appear also to be the origin of space and of time as the expression of the separation of points. The information conveyed by our physical senses tells us that we seem to exist in locations that we call physical bodies.

If consciousness is the progenitor of time and space and whole physical universes, then how does it come to be encapsulated within these little parcels of flesh we identify with (our bodies)?

Could it be that it was the choice to be separate that made us different from the original limitless state from which we came? The only way to become different from it would have been to become limited and specific, and it would then be these specific identities that demarcate our difference from each other and from the state of perfect union of all being from which we came.

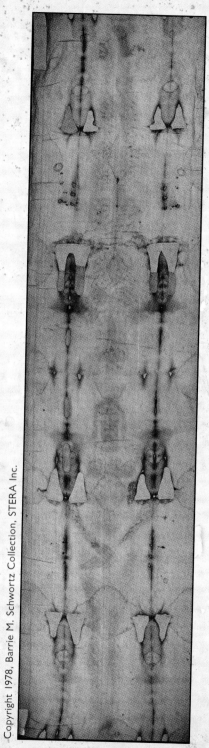

Plate 1. A full-length color photograph of the Shroud of Turin.

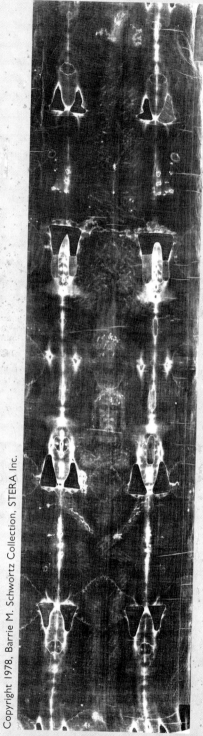

Plate 2. A full-length photographic negative of the Shroud of Turin demonstrating that the body-image negative itself looks like a photograph.

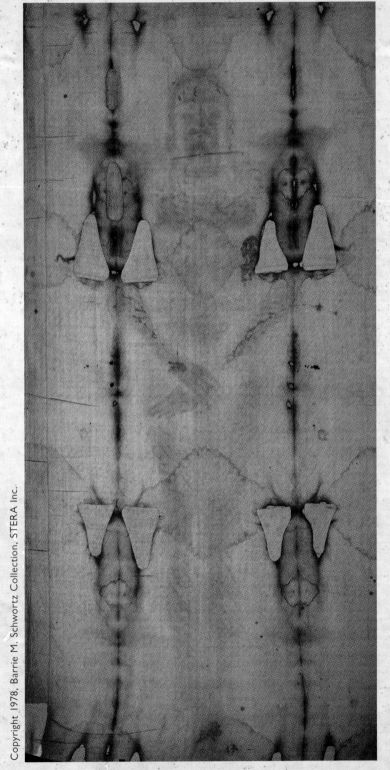

Plate 3. A color photograph of the Shroud of Turin showing the bloodstains and body image from the front of the body.

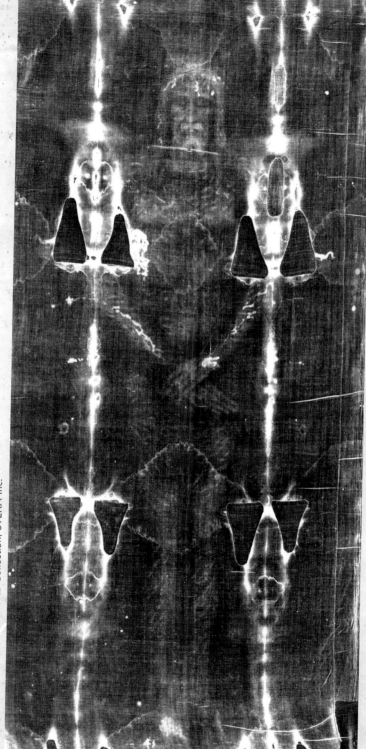

Plate 4. A photographic negative of the part of the Shroud showing the frontal body image and bloodstains.

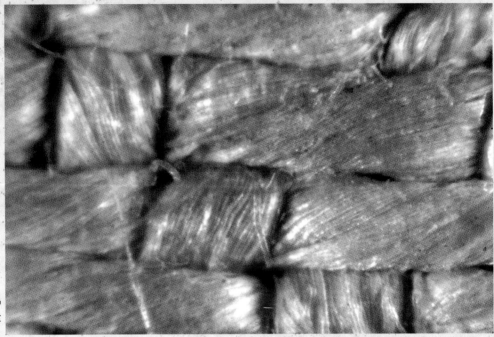

Plate 5. A photomicrograph of an image area of the Shroud demonstrating that the image appears due to a sepia discoloration of some of the surface fibers, similar to the discoloration of paper exposed to sunlight.

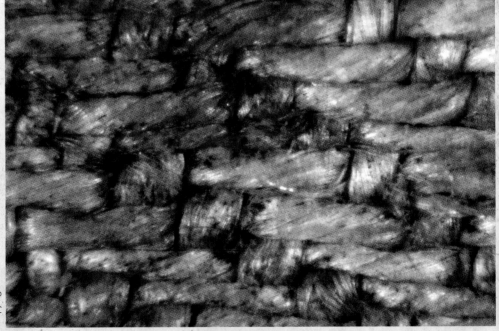

Plate 6. A photomicrograph of a bloodstained area of the Shroud showing that, in contrast to the body image, the discoloration here is due to a substance (blood) that is adherent to the cloth.

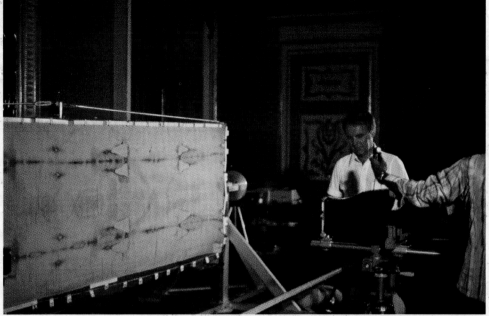

Plate 7. Vern Miller takes photographs of the Shroud.

Plate 8. STuRP team member Ray Rogers examines
the Shroud with a magnifier.

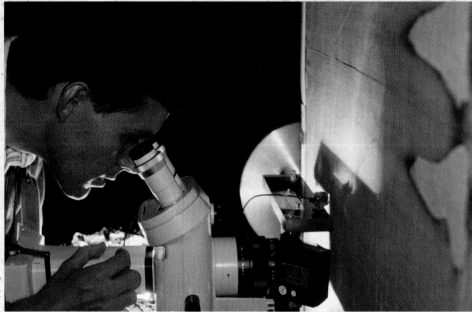

Plate 9. Scientist and STuRP cofounder John Jackson examines the Shroud through a microscope that also had a built-in camera.

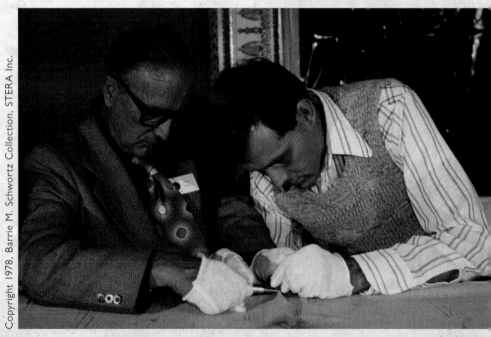

Plate 10. Scientists take measurements of the Shroud.

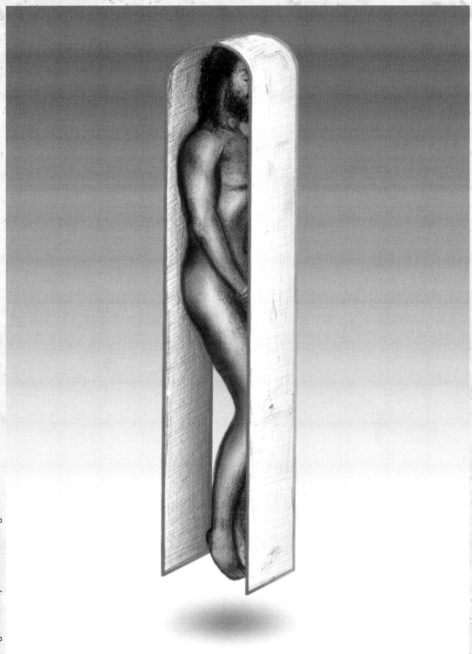

Plate 11. A depiction of the position of the body and of the cloth, based on the evidence suggesting that the man of the Shroud was vertical at the moment of image formation.

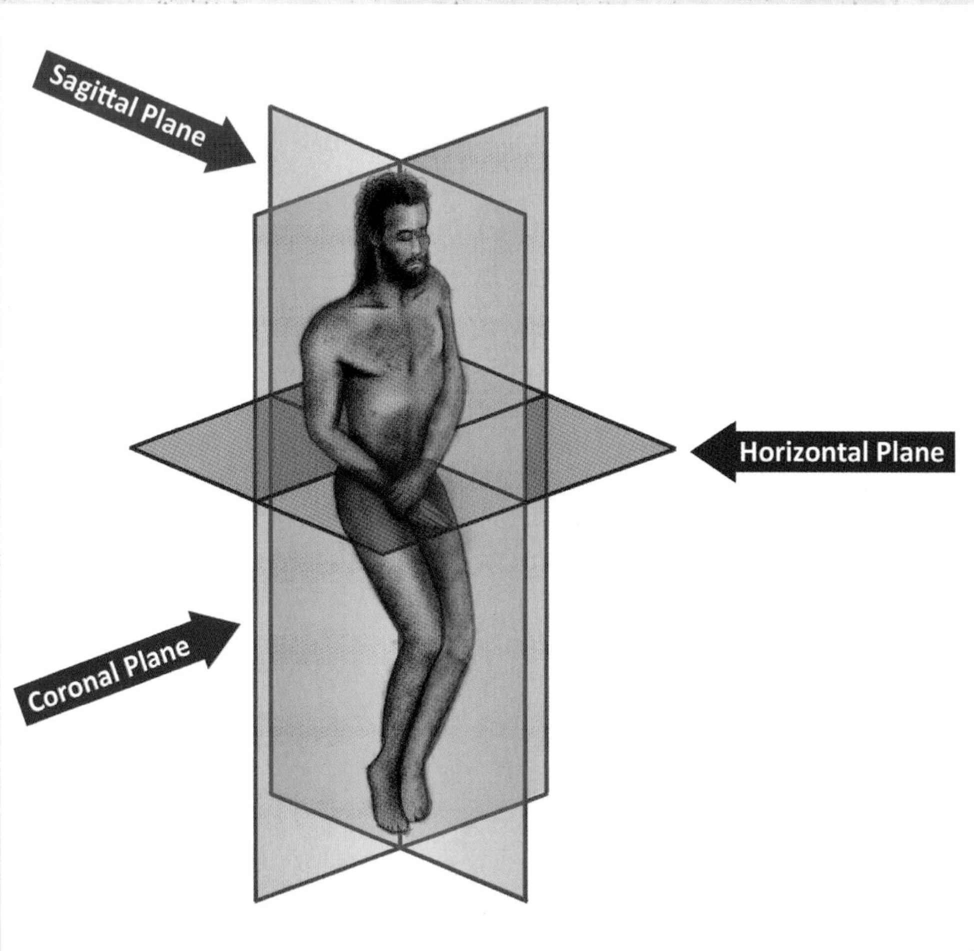
Sagittal Plane
Horizontal Plane
Coronal Plane

Image courtesy of Daniel Langsman

Plate 12. A depiction of the three anatomical planes of the body of the man of the Shroud.

You Have
No Beginning
or End

In this book, I have considered space, time, and matter as simply the products of separation occurring at the big bang. I have suggested that perhaps the instigator of this could have been something that I have referred to as the *"choice of sentient beings to experience separation."* Does this involve a contradiction in terms? If sentient beings choose to exist in separation, then how is the choice made before the separation has occurred? In other words, how could you have made a choice to be you if before the choice was made there was no you as an independent separate entity?

I suggest that this apparent paradox arises because we have a tendency to superimpose our temporal experience onto a state of being in which it does not apply, for example, when we use words like *before*. The state of no'thing'ness from which the physical universe emerged is a state in which there is no space, time, or matter. As there is no time, there is no change, but also it is not static like a rock or a statue, as these objects are seen as static only within a frame of reference in which there is time. They are the *objects* of our *subjective* perception. It is *we* who *perceive* them to be unchanging. Time, then, could be seen as resulting from the separation between and difference between subject and object.

We are used to thinking of subject and object as being distinct. *The existence of empathy perhaps is a clue to the route map for undoing this separation between subject and object.* If you care about other sentient beings, then you might feel as though something that happens to them has happened to you. This is perhaps because you sympathize with their pain or take pleasure in their happiness, for example. If we imagine this distilled, as it were, to its purest form, then subject and object become identical, and there would then be no such thing as distinct objects. All subjects of perception would be one and arguably limitless. As an exercise, perhaps you could think of occasions when you have felt this—when you have been moved by sympathy for someone's suffering and also when you have taken pleasure in someone's happiness. It is interesting that in many near-death experiences (discussed in more detail later in this chapter), when people experience a "life review" (which is probably where the phrase "seeing your whole life flash before you" comes from), many people report that the whole experience is made up of moments of empathy. People experience their interactions with others from the other person's perspective, so that if they were kind to someone they feel as though the kindness was done to themselves, or if they harmed someone they experience this as the harm being done to themselves. Seen in this way, a lot of the teachings of people like Buddha and Jesus don't need to have connotations of institutional religions but seem instead to be simply rational expressions of good sense.

Physical universes, then, are arenas in which subject and object divide. Remember, though, that the source state of no'thing'ness is beyond space and time and so does not change. It does not "become" the physical universe, as to do so it would have to change into it. It does not change, though, as it is not within what we call "time." Also, it does not "create" the physical universe, as to do so it would be performing a temporal action and would therefore, by implication, have to change. This is because the actualizing of the option to "create" would define it as having become different from how it was when it had not actualized that option "yet."

As we have seen, Schrödinger suggested that time is a product of mind. He also derived from this that mind cannot have a beginning or end. This means that we (that is, you and I and all other sentient beings) trace our origins from the ultimate state of union—of no'thing'ness. I use the term *no'thing'ness* to express the notion that you are "someone" rather than "something." Although your mind may perceive things, the center of your awareness, your *mind's eye*, if you like, is itself invisible, intangible, and not material. It is you and you are not a thing. The physical universe itself could also have arisen as a result of "our" separation from this state and therefore from each other. I have argued that whereas force can be a director of change, its absence provides for the potentiality of *options* and that the complete absence of force in the original state of no'thing'ness provides the limitless potentiality of all options through freedom of will. The availment of choice in actuality defines the chooser in terms of separation from the prior state of being "the all," and in producing separation, space, time, and matter arise. These are the outward expressions of this separation.

Rather than being propelled into this universe, we have actually chosen to exist in separation, and in so doing have created and continue to reify or "make real" the arena in which we exist. This would of course be consistent with what quantum physicists such as Eugene Wigner have said about consciousness being necessary for quantum mechanics. So consciousness is necessary for quantum mechanical processes to exist, and quantum mechanics is necessary for the physical universe to exist.

Is it then so unreasonable to postulate that consciousness was instrumental in causing the big bang?

We have explored the nature of consciousness and considered the notion that it consists of awareness and will, where *will* implies the potential for independent causation. So, if consciousness caused the big bang, then it would seem that this big bang would have been the result of the action of will. The big bang is the starting point of space and time, and therefore it is the starting point of separation, so I would suggest that if the cause of the big bang is *choice,* then it was a choice *for* separation. The same choice that is responsible for the separation of points, which we call space, and the apparent separation between you and me, which we call individuality.

That which we commonly call "the universe" is a state of separation between points. It is generally seen as being defined in terms of matter, space, and time. Matter exists as separated locations of force that, on a large scale, are moving apart at an accelerating rate through universal expansion. Observing this universal expansion led cosmologists to realize that there was an implied boundary condition inasmuch as it would seem that the entire observable universe must have begun as a single point that had the properties of a singularity. Matter follows the second law of thermodynamics and always tends toward increasing states of disorder. This also implies that there would have been a highly ordered starting point. We have also discussed evidence that matter itself is a derivative of consciousness. I couldn't put this better than Max Planck, the founder of quantum theory, did as quoted in the *Observer* (a major British newspaper) in 1931:

> I regard consciousness as fundamental. I regard matter as derivative from consciousness. We cannot get behind consciousness. Everything that we talk about, everything that we regard as existing, postulates consciousness.[44]

If consciousness or sentience lies at the root of existence itself, then how would we characterize what the properties of this consciousness would be within the state of no'thing'ness from which the physical universe emerged? Again, I use the term *no'thing'ness* rather than *nothingness* as I do not mean to imply a "void." I am referring to the absence of isolated "things" where a thing is seen as an "object." In the state of no'thing'ness, I would suggest that subject and object are one and the same and consist of pure awareness. It is from this awareness that our minds and identities arise as individuality, but within the "ground" of that awareness, we are all one. So one property that we can deduce is that it must have a property of unity because within its realm there is no space or time in which separation could exist. We have also seen that Schrödinger, for example, pointed out that fundamentally there is an implied unity between all sentient beings, even in the separated physical universe. He also made a case that the "now" and therefore time itself are derivatives of the mind and that as time is a product of mind it follows that mind does not begin or end in time and is therefore eternal. So, could there then be two aspects of mind?

The more familiar one to us is, of course, the one in which mind takes the form of separate individuals. There is, within this state of individuality, an inherent implication that our separation, one from the other, is contingent on the initial separation that we call the big bang, which perhaps, as I have said, is itself the expression of the choice to be separate.

The other aspect of mind is perhaps one beyond space, time, and matter in which there is no separation and we are all one! Being beyond time implies that there would be neither change nor stasis. This is because the notion of something being fixed or static implies the passage of time during which mind is seen to be unchanging. If it is the source of all sentience, then it must contain all potentials within it. In other words, you and I and all other sentient beings who have ever lived or existed are in a meaningful way present within this state of pure existence as well as being present or having been present within this physical universe as separate identities.

Considering the belief that many people have that a god (God) cre-ated the universe, it is noteworthy that the man whose image we see on the Shroud (if we are to be convinced by all the evidence that indicates that he was Jesus of Nazareth) is reported to have said, "Is it not written . . . ye are gods?" The problem with words like *god* (or "God") is that they can be said to have as many meanings as there are people who use these words. With that in mind, I feel I owe you an explanation of what I mean when I use this word in this book. The way I try to understand such matters is by thinking with my "heart." So I pay attention to how I feel about those people I care about and in whose presence I learn about "peace" and what it means to be "whole," and then I try to consider how I might be if I could be for everyone what they are to me. This, of course, brings it home to me how so very far away I am from being such an embodiment of compassion. However, it also informs me that I, you, and all other people have the capacity to be such as this. "God" is the realization of our full potential. It is what each of us were, are, and could be without our imperfections. As it is our imperfections that define the separation of our identities, it follows that if we were perfect, there would be no distinction between you and others. But you would not be losing your identity. You would be regaining it! The problem is that we tend to identify our "selves" by what divides us and makes us distinct. This tends to make us lose sight of the fact that although the contents of our awareness may differ, the fundamental ground of that awareness is one and the same in all of us. When you forget your "self" in your unconditional love for others, you are not extinguishing your existence, you are realizing your existence.

I think the comment of Linde bears repeating here in the context of the nature of humanity and indeed all sentient being and how we relate to the physical universe:

Without introducing an observer, we have a dead universe, which does not evolve in time. Does this mean that an observer is simulta-neously a creator?[3]

If, for the sake of argument, we were to consider the possibility of the universe occurring due to "our" choice to exist in separation, then why would such a decision have been made? I have already expressed my view that, by definition, if there is such a thing as free will, then there can never be a complete and final explanation for why a particular decision is made. This is because if there were such an explanation, then it could not have been otherwise, which would imply that it was predetermined and therefore not free will.

"Hang on a second . . ." some may ask again. *"How could I have been a part of the decision to create the universe considering that I only came on the scene several billion years after the big bang?"* It seems to me that this question implies the supposition that "you" began when you were born (or conceived) *this time around.* My contention is that if people like Planck, Schrödinger, Eddington, and others were right, that matter is a property of mind rather than mind being a derivative of matter, then it would seem to follow that what we call "you" or "me" cannot be simply an emergent property of the arrangement of atoms or cells in our brains.

I would not dispute that we appear to be viewing the world from behind our eyes or making physical actions in the world through decisions that involve the use of our bodies. However, this is not proof that we are the products of these bodies.

If you turn on your radio and hear your favorite song, you could be forgiven for laughing if someone told you that they believed that the radio was singing to you or that the radio was the songwriter.

INFORMATION AND KNOWLEDGE

What is knowledge, and does it differ from information?

A computer can be programmed with *information,* but it is not *aware* of that information and therefore cannot *know.*

Would it be fair then to surmise that *knowledge* does not come through *information* alone but requires *awareness*?

There are many reports in the medical and scientific peer-reviewed literature of events known as near-death experiences. These usually occur to people who have had a cardiac arrest and then been successfully resuscitated. During the cardiac arrest, the brain was not in a fit state to lay down new memories, and yet after these people have been resuscitated, they recount their experiences during the arrest in great detail in a clear, coherent, and consistent manner.[45] This, in itself, has led some scientific researchers, notably including cardiologist Dr. Pim Van Lommel,[46] anesthetist Dr. Sam Parnia,[47] and neuropsychiatrist Dr. Peter Fenwick,[48] to conclude that the mind cannot be wholly dependent on the brain as experience appears to be possible without a functioning brain.

Some neurobiologists have argued that these experiences are artifacts of activity in parts of the brain deprived of oxygen, but Fenwick reminds them of evidence that brain function ceases during cardiac arrests. Research has indicated that around 10–20 percent of people resuscitated from cardiac arrests recount memories that are consistent with typical near-death experiences. Many of these, which Fenwick refers to as *veridical* near-death experiences,

> include an ability to "see" and recall specific detailed descriptions of the resuscitation, as verified by resuscitation staff.[48]

Many people who have had near-death experiences perceive a brilliant light. Their descriptions of it share a recurring theme that in the light was all knowledge and perfect "love." Being near to this light reportedly made them feel so at home that they didn't want to leave it, even to the extent that mothers with young children wanted to abandon them to be with the light rather than return to the lifeless body they left behind in the resuscitation room or operating theater.

When asked about what they can remember of the knowledge that was "in the light," people don't seem to recall any specific *facts* or *information,* but the light seems somehow to be associated with a sense of immense love and happiness.

It would seem that what they glimpsed was recognized and known by them in an instant as *truth* without explanation or derivation. This insight was associated with a sense of unconditional compassionate love. To quote Fenwick:

> This combination of radiant light, wisdom and compassion correspond precisely to the descriptions given.[48]

Not all near-death experiences are pleasant or heartwarming. However, a common feature of many positive and negative near-death experiences is that following them, the near-death experiencers find an inner certainty that death is a transition rather than an end and also find that their value systems are overhauled as they become deeply convinced of the value of all humanity to the extent that in many instances they devote their lives to the service of others. If reincarnation is true (see pages 56–57), then perhaps most or all of us have had near-death experiences (and death experiences) before. The near-death experiencers differ from the rest of us merely in that they have conscious recollections of the experience.

The light that people experience during many near-death experiences, I would suggest, is a clue to the difference between knowledge and information. However, even information is dependent ultimately on sentient beings to make it information. A book is just patterns of ink on paper until someone reads it. A lullaby is just vibrations of molecules in the air unless someone is soothed by it. Information is a broken, diffracted derivative of knowledge, and yet even it only exists if it can be known. So knowledge is fundamental. But you could know all the positions of all the grains of sand on a beach and that wouldn't add one iota to the meaning of your memories of walking on it. You could know all the electrical activity in the brain of a loved one and in your own brain and it wouldn't give you any clue about what that person means to you.

Quantum theory shows us that the mind of an observer does more than subjectively give meaning and interpretation to the observed; it actually makes the observed real.

We as observers are characterized by the capacity to know, to be aware, and everything else; time, space, and matter seem to spring from this as products of our observation process.

If our state of being aware as observers is enough to make the world real, then why are there so many of us? What is individuality?

What if the natural state of all being *is* the light seen, felt, and known by near-death experiencers? Might it follow from that that our individual identities are simply measures and measurers of our specific patterns of ignorance or "darkness" as referenced to that light?

We would then exist as an amalgam of light and darkness. Selfishness and ego would simply be announcements of this darkness or "shadow" within us where we have not allowed the light through.

People with more empathy and compassion would simply be those who function more out of the "light" than the "shadow." If the light is common to all sentient beings, then the distinction between self and others would become less important to them. The statements by people like Jesus of Nazareth that their most important advice is that we love our neighbor *as ourselves* could then be seen as the ultimate in reason and good sense if ultimately and originally we are all one.

I should clarify that when I refer to the light here, I am not using the term in a poetic or metaphorical sense, but I am suggesting that there may be an actual phenomenon involved in all living beings in various states of diminution and distortion. This light perhaps reaches its fullest, most complete expression with the light that people experience in near-death experiences and the light that formed the image on the Shroud, but nevertheless it is the same light that is present within each of us to varying degrees according to how we live and the choices that we make.

So is it through the light that we exist as sentient beings? Is it then the darkness that limits us and holds us bound in ignorance and separation? If so, then surely the most important *information* we can receive concerns the preciousness of this light, wherever it may be, and therefore the limitless and equal value of ourselves and others.

So what, if any, is the relationship between this light that is perceived in near-death experiences and the light that would seem to have been responsible for the image on the Shroud?

What if they were one and the same?

Remember, Einstein's equations tell us that from the perspective or reference frame of light itself, there is no separation and that the origin and destination of the light ray are united in space and in time. This would imply that, in a very real sense, space and time themselves are properties that emerge from the "freezing" of light into matter.

It is interesting that the man of the Shroud (again assuming for the sake of argument that we accept the considerable evidence that he was Jesus of Nazareth) is reported to have said, *"I am the light."*

I have suggested that we are all amalgams of light and shadow and that it is the shadow that marks us out as discrete, separate individuals in physicality. If this is correct, then perhaps the event that formed the image on the Shroud is an example of what can happen if a human being completely divests himself or herself of all "shadow" and achieves one's full potential. In view of the arguments that suggest that the root of our existence as sentient beings is a state of unity from which we all emerged, this would be consistent with a life lived truly loving one's neighbor as oneself with total dedication and magnanimity.

Perhaps he demonstrated how it is possible for human beings truly to *become* the light that people perceive during NDEs.

If this is so, then the image on the Shroud would really be a physical record of a direct interaction between this light and the physical world. The light became a man, and that man became the light. I'm sorry if this sounds like a platitude, but I hope that, taken in the context of what I have written in this book, you will see that this is not meant in a religious sense. I hope that you will also see from what I have written that I have made a case that all human beings originally derive from that same light.

If we are to attempt to achieve a unified understanding of ourselves, each other, and the world we live in, then we need to consider the nature of mind and matter and, therefore, of subject and object. In this sense, when I refer to subjectivity and objectivity below, I do so not in the sense of the difference between subjective opinion and objective reality but in the sense of the difference between the subject of perception (i.e., one who perceives) and the objects or contents of that perception.

SUBJECTIVITY AND OBJECTIVITY

Surely, a scientific attempt to understand human consciousness and its potential has to extend to a consideration of the interface where subjectivity and objectivity meet. In other words, to try to understand the connection between mind and matter must occur.

It is arguable that the "Western" scientific culture and, in particular, the empirical method can sometimes stunt our thinking when we try to understand the mind. For example, consider the saying, "Cogito ergo sum" (I think, therefore I am).

While considering those words, we can observe the fact that we are thinking and logically deduce that there is something or an entity that is doing that thinking. We identify that thing or entity as "I" and deduce that that "I" exists.

Is this a scientific approach?

As all scientists know, the word *science* literally and etymologically refers to a field of study that pertains to knowledge. The scientific method is often described as an attempt to understand or derive knowledge about the world through developing theories that can be informed by and tested by reproducible observation. However, many people equate the *scientific* method with the *empirical* method, which is actually subtly different.

The empirical method is specifically concerned with information gathered by the senses. This would mean that "I think, therefore I am" could be scientifically tested but could not be empirically tested. As thoughts are invisible, silent, odorless, tasteless, and intangible, then is it the case that we cannot know scientifically that we exist?

Many might say that here I am straying from science into philosophy, but if quantum theory (which is mathematically the most accurate and verifiable branch of science) depends on the existence of a conscious observer to collapse the wave equation and make reality real, then how can it be of no relevance to science whether or not we truly exist as conscious observers?

When I have had discussions with people about consciousness and the mind, I have often heard people say that consciousness is the result of self-awareness and that this is a result of the evolution of *language,* that we became self-aware once we had *words* to describe self-awareness. Again, I take issue with this because it seems to me that there is a clear distinction between the *state* of being aware and the *content* of that awareness.

A simple illustration of this is that the term *self-awareness* exists

on this page that you are reading, but does that really give this piece
of paper a mind or make it conscious? Computers also can contain
language encoded both in their software and in information processed by them, but does that make computers conscious? Proponents
of artificial intelligence might answer in the affirmative, but would
they seriously argue that it should be considered an act of murder to
unplug a computer? It's interesting that computer science began in
large part as a result of the ideas of Kurt Gödel, who was a friend
and contemporary of Albert Einstein at the Institute for Advanced
Study. Gödel was a mathematical genius who is most famous for his
two incompleteness theorems. The first of these theorems was used
notably by Penrose as a major part of his argument against the possibility of true *artificial intelligence*.[49] In a nutshell, Gödel's first
incompleteness theorem demonstrates that no computational method
(of algorithms based on axioms) is capable of deriving all the simple
arithmetical "truths" about numbers. Gödel argued that mathematicians rely on intuition in order to see the consistency of mathematical proofs. This intuition can never, even in theory, be reduced to an
algorithm or a computer program!

Many theoretical physicists have spoken about being awestruck by
the capability of mathematics to describe the physical universe. This
raises the question of how matter, space, and time "know" that they
should follow the equations. Galileo wrote that "[the universe] is written in the language of mathematics." The cosmologist Professor Sir
James Jeans famously said that God is a mathematician.

If mathematics is reliant on mind for its "noncomputable" realization
and matter is a derivative of mind, then perhaps it is not surprising
that mathematics works in describing the universe!

The enigma of why the physical universe is understandable with
mathematics can be seen in the terms of this book to be soluble in
the same way as the mystery of why the universe began in a state of
order rather than chaos. Mathematics is a product of reason. Reason

is a product of simplicity and symmetry. These are both products of order. Order is also a property of no'thing'ness. No'thing'ness is a product of peace (i.e., the absence of force). Peace is a product of unity. Unity is the property of that in which all absolutes are contained.

The Clues
on the Cloth

We have seen in chapter 1 that forensic scientists have been able to piece together some of the events of the last few hours up to and including the death of the man on the Turin Shroud from studying the bloodstains and serum stains on the cloth. (Serum is the liquid residue from blood that separates from the clot when blood coagulates.) We have also seen that the formation of the bloodstains happened by a simple contact process that occurred when the recently deceased body was wrapped in the Shroud. From a scientific study of the characteristics of the image on the Shroud, we have seen that the image-formation process on the cloth happened after the blood-stains had formed and after the bodily death of the man whom it wrapped.

Just as there are forensic clues in the bloodstains, there are also clues in the image regarding the event that took place when the image formed. These clues have only become apparent through the application of modern technology to the Shroud and would not have been apparent in medieval times, for example. It was only with the development of photography that the photographic negative properties of the image could be discovered and only with twentieth-century technology that the distance-coding properties of the image were revealed. Both the image and the bloodstains have been shown to provide detailed and accurate anatomical and pathological information way beyond that which was known in medieval times, and again this information is only apparent to us now using modern techniques of investigation and analysis. Now might be a good time to reconsider the evidence on the cloth in the light of what has been discussed in previous chapters about the nature of mind and what modern science (quantum theory) implies regarding the connection between mind and matter.

THE IMAGE ON THE SHROUD

Dr. Gilbert Lavoie has presented evidence that the bloodstains occurred when the Shroud wrapped a recumbent corpse.[14,50] He also demonstrated that the body image on the Shroud is more consistent with that of an upright man! More details of his work are discussed in the appendix. The lack of flattening of the calves and buttocks on the dorsal image supports the postulation that the man of the Shroud was vertical when the image formed, as does the fact that the hair appears to be hanging down on the shoulders rather than spread out behind the head and neck. The position of the feet (see plate 4), with the toes of the right foot pointing downward and the left knee appearing to be slightly more flexed, suggests that the body was not in a standing position but intriguingly appears to have been suspended above the ground. One possible explanation for this would be if the mass of the body might have somehow become less. If this were the case, then how might this have occurred?

I have presented a case that mind and matter are a continuum and that as living human beings we are an amalgam of both. As living conscious sentient beings, we *are* the mind-matter continuum. That which looks out at the world from behind your eyes is the power of all power. Without such as this, there would be no physical universe, and through such as this lies the potential for you to achieve a status of existence compared to which all the billions of galaxies are as nothing.

I have also suggested that the origin of matter itself was the result of a choice for separation from an original singularity and that matter is simply a localization of force. If mind and matter are a continuum, then so too are freedom of will and force. This is because these represent what I have called primary and secondary causation, respectively, such that a free choice is an "uncaused cause," whereas an influence of force is itself the result of defined causes and conditions. I have argued that it is our limitations or restrictions of mind that differentiate us from each other and from the original unity of awareness from which we emerged. These restrictions are expressed through thought and actions that are divisive, such as selfishness, arrogance, racism, and so on. I have suggested that it is the reduction in power of the "unenforcedness" or "peace" within us that is responsible for our existence within physicality and ultimately for the physical universe itself. If this is so, then it would make sense that if someone lives a life that consistently unites with others through compassion and recognition of the fundamental union of all sentient beings, then force could be reduced in the location in which that person's will predominates (i.e., his physical body).

Einstein's equations show that mass is responsible for the curvature (or distortion) of space and time. Mass, in turn, is seen as the condensation of energy. This describes the latent enforcement within matter.

I should clarify that when referring to mass, I am of course not referring to the amount of matter within a human body. I am clearly not suggesting that people who are of slight build have less restricted minds than others! What I am suggesting is that the atoms in the body of the man of the Shroud may themselves have become lighter, and also that the distortion of space and time was reduced in the immediate vicinity of the body of the man in the Shroud at the moment of image formation.

The implications of this hypothesis would be that if the mass of his body had reduced to the extent that it had around the same density as the air around it, then this might have been what caused it to be suspended above the ground, as depicted in plate 11.

PEACE BEHIND THE "I"S

If we consider what Bohm explained about matter being frozen light, then perhaps the freezing of light as matter is a result of the choice of sentient beings to separate (see chapter 5).[42] I have made a case that force is what will becomes when will is no longer free and therefore that it is our restrictions and limitations of mind that hold us bound in this universe of separating parts. If the man on the Shroud was Jesus of Nazareth, then it is interesting that he seems frequently to have referred to the limitless potentiality latent within all human beings. For example, he is reported to have said that "nothing will be impossible unto you" and that those whom he referred to as "peacemakers" would be called "children of God."

The term *peacemaker* is an interesting one. The word *peace,* or *shalom,* is much repeated in the Hebrew scriptures, for example, in Isaiah 9:6, it is said that there would one day be a man who would be known as "Sar Shalom," or "the Prince of Peace." *Shalom* means "peace, completeness, wholeness, and well-being." Does it not seem reasonable that when Jesus also spoke of *peace,* he was referring to the absence of *force* and the underlying union of neighbor with self? These two would be correlated if it is our restrictions that define us as separate individuals. If freedom is the property of the absence of force (see page 85), then it would make sense that if it were possible for a human being to lose all restriction of mind, then this could have a direct effect on that person's physical body too. There are reports, for example, of Jesus walking on water and of certain individuals in history, such as Siddhartha Gautama (the Buddha), rising above the ground. In my thesis, this is all consistent with physics once the relationship between mind and matter that I have referred to as the mind-matter continuum is understood.

One clue to this might be found in the anecdotal report of Peter walking on the water (Matthew 14). The story as reported suggests that Peter began to sink due to his attitude of mind, which is referred to as "fear" or "doubt." It is interesting that if this report were accurate, then there is an inherent implication that thought might under certain circumstances have a direct effect on the "inertial enforcement" of matter.

If for the sake of argument we imagine that the image on the Turin Shroud might be a photographic negative image of a resurrection event, with the image resulting from a sudden burst of radiant light emanating from the body that the Shroud had wrapped, then what would the implications of this be for our understanding of the nature of life from a physics perspective?

Despite numerous attempts, nobody has been able to manufacture a living organism de novo. All that has been achieved is the production of chemicals such as amino acids, which are often described as the building blocks of life, as though it is just a matter of time before someone finds a way of assembling these to make a living cell.

In 2010, Craig Venter, founder of the J. Craig Venter Institute in Rockville, Maryland, claimed to have made synthetic life. What his team had actually done was to implant biosynthetic DNA into the cytoplasm of a natural cell. This is in no way close to synthesizing a cell from scratch as it relies on using natural life as its foundation. I would suggest that there will never be such a thing as artificial life and that all life traces its origins through continuous lines leading right back to the origin of the physical universe itself.

Some people will no doubt think that it is preposterous to suggest that life exists as a continuous line right back to the big bang. They will argue that throughout much of the history of the universe, there were no atomic elements other than hydrogen and helium and also that in the very early universe, there were not even atoms at all. They will point out that life as we know it is carbon based and that there was no carbon at this early phase of the physical universe. However, this might presuppose that we know what form life must take, and there is always the danger of being anthropomorphic in these matters. If, as I have argued, the brain is a transducer for consciousness rather than its creator, then how can we be sure that there wasn't once a hydrogen-based living being? This may seem a strange notion, that life could prevail in something so insubstantial and uncomplicated, but the evidence from near-death experiences, in which consciousness appears to persist beyond the body while the brain is not functioning, would suggest that mind does not require substance at all. The other point is that quantum mechanics demonstrates that substance itself is brought into existence by conscious observers. This of course strengthens the case that consciousness and not substance is primal and fundamental.

I have presented a case in this book that the state of no'thing'ness from which the physical universe emerged exists beyond time and space, as time and space begin with the physical universe. The currently accepted laws of physics do not account for how there can be a "*now*" or how what we perceive to be a *flow* of time comes to exist (see chapters 5, 8, and 9). Schrödinger, as I have said, derived from this the idea that time is created by mind and therefore that mind has no beginning or end. This places all of us as sentient beings right back at that original singularity. This is not meant in the sense of our physical bodies (as they are now) being there, but of our streams of consciousness or awareness having their (temporal) origin there. However, beyond that singularity, there is no space and no time and therefore no separation, so we must all be one within that pluripotent state. Also, as it is beyond time, it would mean that we did not begin. Only our separation and our restrictions began.

My suggestion is that physical life is simply an amalgam of the products of that original state of existence with matter in the form of physical organisms.

Perhaps physical life derives from a continuous line leading right back through the initial universal singularity, and it is our bodies' antecedental or ancestral connection to this line that makes them capable of sustaining life.

How might this relate to the event that occurred in a tomb in Judaea to form the image on the Shroud? Forensically, from studying the bloodstains, we see that the man of the Shroud was already dead when he was wrapped in it. As there is no image under the bloodstains, it appears that the formation of the image occurred *after* the blood stained the fabric. If, as the evidence suggests, his dead body did indeed rise up into the vertical position so that it was suspended above the ground and then emitted a momentary, intense burst of radiant energy, then what happened next?[14,50] Could it be that his body was transformed in such a way that he was able to bring it back to life?

Perhaps in being an embodiment of his teachings and living the life that he did, it might make sense to deduce that the man whose image we see on the Shroud came close enough to the primal state of union that I have referred to from which all potentiality springs that he was also able to bring life to inanimate matter. As that body in the Shroud held all the information about how he had previously intermeshed with the material world, if he rejoined the physical world, then it would seem to make sense that it would be that body that would come back to life.

It is interesting to consider the sequence of events involved in the formation of the Shroud image. If we consider the forensic evidence on the Shroud, then this suggests that at the moment the image formed, the body was upright and suspended above the ground. Once the body was in that position, it would seem from the physical evidence of the Shroud image that there was a sudden, momentary burst of radiant energy from the body in the Shroud that caused the image to form. How do we make sense of this sequence within an attempt at achieving a unified understanding of mind, space, matter, and time?

How would a reduction of the mass of the body to the extent that the body had the same density as air be possible, and can this still be consistent with the principle of conservation of energy? To understand this we should remember that Einstein showed with the equation $E = mc^2$ that energy and mass are equivalent so that what is actually conserved is mass-energy. Hawking pointed out that the universe is "the ultimate free lunch" as the total energy of the universe is and has always been zero![31] The mass-energy of all of the contents of the universe is precisely offset by the negative energy that is present due to the gravitational pull that results from the separation of matter in the universe. If this gravitational pull were also reduced minutely by this event, then this principle of conservation of energy would still apply.

A DIRECTIONAL BURST OF RADIANT ENERGY?

If the image on the Turin Shroud was indeed formed as a result of a burst of radiant energy, then we should be able to do some detective work to find information about the characteristics of this burst by studying the image. The first thing that we find is that it would have to have been *directional*. If the body had "shone" equally in all directions, then there would have been no clearly defined image of the man on the cloth. To recap, we can consider the image on the Shroud (plate 11) by visualizing the cloth wrapped over the top of the head and hanging down over the front and back of the body. I have been to Turin to three of the expositions of the Shroud. When you look at the Shroud as it is displayed, stretched out as a flat sheet, you can see the image of the front of the body and of the back. There is no image in between the two where the cloth would have been over the top of his head when the image formed. There is a water stain between the two images but no image of the top of the head. If you imagine a hypothetical burst of radiation from the body occurring in all directions, then there would be image formation at the top of the head, and also the image would have been very diffuse as it would have been expected to result from radiation spread out all over the cloth.

It would therefore seem that if the image was formed by a radiant burst from the body, the burst would have to have been projected in one axis only, forward and backward, as it were. The physicist John Jackson has postulated a hypothesis to explain this that suggests that the recumbent corpse dematerialized, producing radiation as it did so, and that the cloth that had been covering the front of the body collapsed downward as a result of gravity through the space where the body had been and that the image was imprinted on it as it collapsed through the resulting field of radiant energy.[51] This would be consistent with the directional nature of the radiation, although, as I have stated elsewhere in this book, there is further evidence that suggests that at the moment of image formation, the body was actually upright rather than supine.[14,50]

One criticism that is often mentioned about the hypothesis of directional radiation concerns the distance coding in the image. This distance coding was first discovered by Jackson in the 1970s; this means that the intensity of the image depends on the distance there would have been between the cloth and the body, so that, for example, parts of the body that would have been closer to the cloth, such as the nose and the hands, have a higher intensity of image formation than parts that would have been farther away.[52] This is consistent with something known as the inverse square law, which describes how radiation intensity reduces as the distance from the source increases. However, the inverse square law usually depends on the assumption that radiation is spread out in all directions so that it becomes weaker the farther you are away from the source as it becomes more spread out. If the radiation had not been spread out, then there would normally be no reason for it to reduce in intensity, in this case as the distance between the body and the cloth increased. So, the argument goes, one should expect either distance coding and a diffuse image spread all over the cloth or no distance coding if the image is only projected in one axis.

So how can it have both directional properties *and* distance coding?

The ENEA experiments of physicist Paolo Di Lazzaro and his colleagues used vacuum ultraviolet radiation.[20] What this means is that the short wavelengths that were used would not be transmissible through the air and would have been absorbed by the air before they reached the cloth. However, the wavelength the Italian scientists used was near the edge of the vacuum ultraviolet range. Wavelengths slightly shorter would have been partially absorbed by the air. The reason they couldn't try the experiment with slightly shorter wavelengths was that there are no commercially available ultraviolet laser sources with the relevant wavelengths of emitted light.

This is where I am in a position to put forward a suggestion, one that scientifically I may call a hypothesis. The ultraviolet light that formed the image on the Shroud could have been of a slightly shorter wavelength than the light used in the ENEA experiments. If so, then it may have been partially absorbed by the air between the body and the cloth, and this could account for the distance coding on the image!

The Italian physicists were able to obtain a Shroud-like discoloration of linen using vacuum ultraviolet radiation that had a wavelength of 193 nanometers.[20] Between 200 and 150 nanometers, ultraviolet radiation is absorbed by oxygen in the air, and below 150 nanometers, it is also absorbed by nitrogen.

This may have profound implications for our understanding of the distance-coding properties of the image. For wavelengths above 186 nanometers, there is no significant absorption of the ultraviolet radiation by the air. At a wavelength of 120 nanometers, it is absorbed by the air after only around 0.1 millimeters. The body image appears to have been formed with the cloth within centimeters from the body. There may be a wavelength somewhere between 120 and

186 nanometers at which the exponential absorption of the radiation by the air between the body and the cloth generated the distance coding of the image and also the extinction distance of the image at around 4 centimeters. Di Lazzaro's experiments show that the extinction distance would not have been the distance at which the intensity of the radiation had reached zero but rather the distance at which it fell below the threshold required to alter the chromophore in the surface of the fibrils to cause the discoloration.

There is the suggestion here that further experiments could either support or refute this. If an appropriate laser source can be found, then the ENEA experiments could be repeated at varying distances between the laser source and a piece of linen. However, we still need to account for why the radiation all seems to have occurred along one axis.

To recap, the evidence demonstrated by Lavoie shows that at the moment of image formation, the body of the man on the Shroud was upright.[14,50] This means that the long axis of the body was aligned with the Earth's gravitational field. If the left-right axis of the body had been aligned also with the horizontal component of the Earth's magnetic field (i.e., north-south), then the front-back, or antero-posterior axis of the body would have been aligned in the east-west axis. This alignment of the body with the prevailing ambient force vectors might have been the start of the process of reducing the enforcedness of the space occupied by the body.

It is interesting that before his death, Jesus is said to have predicted that he would return. To quote Matthew 24:27,

> *For as the lightning cometh out of the east, and shineth*
> *even unto the west; so shall also the coming of the*
> *Son of man be."*

What if he was actually predicting his return with that saying and also giving a description of the moment when the Shroud image formed, as the moment of his return *is* the moment that the Shroud image formed?

One point to consider in all this is that it would seem that the burst of radiant energy was a synchronized momentary event that occurred from the whole body all at once, so it would seem that the whole body was acting as a single quantum unit, perhaps in a manner akin to something known as a Bose-Einstein condensate. These states used to be considered to occur only at very low temperatures, but there is intriguing evidence suggesting the existence of similar quantum coherence within biological structures, which would mean that it is feasible that they can occur at environmental temperatures.[53,54] There were two prevailing force vectors acting on the Shrouded man's body, in the up-down (gravity) and north-south (Earth's magnetic field) directions, and perhaps these were reduced to some extent by the changes that were happening to his body, which left one remaining axial direction (east-west) to effect a 360-degree reduction in what we might call the "tension of space" around his body in all directions. Perhaps then, the energy resulting from the undoing of some of the tension implicit in the frozen light that made his atoms might then have been released in that direction. I should clarify that this is of course highly speculative, but it is intended as a postulation to encourage discussion, debate, and further developments leading toward a more comprehensive understanding of the image-formation process. When one considers the evidence presented by Lavoie, which implies that the man of the Shroud appeared to be suspended in the air at the moment of image formation, then it might seem less bizarre to suggest that associated with the process leading to image formation there was also a reduction in the prevailing forces such as gravity or, at least, a reduction in their effect on the body.

If the body had indeed been in something like a Bose-Einstein condensate state, then there could also be some other interesting possibilities regarding the events around the time of the image formation.

The tomb contained the dead body of the man on the Shroud, and it would not seem unreasonable to assume that the body had been laid out supine and therefore horizontally. However, we have seen evidence that at the moment of image formation, it appears as though the body had been suspended vertically in the air.[14,50] This being so, we are left to wonder how it is that the supine body rose up off the slab and rotated through 90 degrees to take on a new orientation. Did the body actually move through all the positions in between being horizontal and being vertical, or did it transform immediately from one position to the other?

We know from quantum mechanics and from empirical observation that at the quantum level, particles such as electrons can begin on one side of an impenetrable barrier and disappear from there and then appear on the other side of the barrier without having moved through the barrier! This process is known as quantum tunneling, and much of our present day technology relies on this process (e.g., the electron microscope), and also, as it is an essential requirement for nuclear fusion in stars from which all the atoms in our bodies (apart from hydrogen) were formed, your body and the book in your hands would not exist were it not for quantum tunneling.[36]

If for the sake of argument we consider the possibility that the body of the man on the Shroud were acting from a Bose-Einstein condensate state, then the body would also have been able to appear in the upright position, having previously been supine on the slab. There is an implication that the atoms of the body would have been entangled in a quantum sense with those of the surrounding air and indeed those of the Shroud itself.

The next step might be that the body as one quantum unit transfigured to a state of less enforcement or a lower energy level, thereby releasing the sudden momentary burst of coherent electromagnetic radiation that formed the image.

Is there a clue in the image on the Shroud to the nature and origins of what we experience as three-dimensional space? Remember that the big bang origin of the physical universe is generally considered to be the origin of space itself. This would imply that the "nothingness" from which the space-time singularity emerged was not spatial and had *zero* dimensions. Some cosmologists have speculated that at the big bang, the universe became three-dimensional by first becoming one-dimensional and then two-dimensional "on the way."[55]

If this is correct, and if the man on the Shroud had achieved a state similar to that initial dimensional interface from which the physical universe emerged, then perhaps this gives us some clue about how the radiant burst of energy was all released in only one axial direction and also how the cloth in front of and behind the man appears to have been very close to being flattened out to be parallel to that which anatomists refer to as the "coronal" plane (see plates 11 and 12).

Existence beyond Time

What Are We?

As living human beings, we experience the world through senses such as sight and hearing and perform actions such as speech and movement using signals generated in the brain to make various muscles move. In the preceding chapters, I have made a case that our perception, even of the information from the senses, goes beyond these senses. This is why we *see* or *hear,* whereas a camera or a sound-recording device merely processes information but does not experience or *know* that information. I have also considered the notions of time and free will and presented evidence that the power of mind over matter is not merely some type of fringe phenomenon but is a part of our everyday lives. Every time we make a choice between two available options, we are exhibiting the power of mind over matter. I have demonstrated that many of the luminaries of science have been convinced that consciousness is not merely an emergent property of an arrangement of atoms but is actually fundamental to existence itself. If that is so, then how do we fit into this worldview as human beings? In this chapter, I will explore this question, starting with a discussion of the properties of the *world around us* and considering the relationship between mind and matter and between *self* and *others.*

PEACE AND FORCELESSNESS

The word *energy* in common parlance is often used in a context very different from its use in physics. In physics terminology, *energy* is a term often used to describe the "capacity to do work," where *work* in turn is defined as "a change in the state or position of an object brought about by the application of a force and involving a movement in the direction of that force."

The Falling Apple

To take a Newtonian example and considering gravity as though it were a force (with due deference to Einstein, who demonstrated that it is not a force), we might consider an apple falling from a tree. The gravitational force between the Earth and the apple pulls them both together, resulting in the apple falling from the tree. The work done to the apple, then, is equal to the force on the apple (i.e., its weight) multiplied by the distance it falls to the ground. In this sense, the work done is equal to the reduction in the *gravitational (potential)* energy and also (ignoring air resistance) equal to its increase in *kinetic (movement)* energy which is related to the speed with which it hits the ground. As the apple drops, gravitational (potential) energy is simply converted into movement (kinetic) energy. Energy is thereby related to force. However, we have seen that the total energy of the universe adds up to zero when the *negative* energy of gravity is taken into account. The universe, then, can be seen to have emerged from a zero-energy no'thing'ness state, and in so doing, it generated, and became an expression of, balanced positive and negative energy, which still add up to zero such that the total energy of the whole universe is conserved at a constant level in accordance with the first law of thermodynamics (known as the law of conservation of energy, which states that the total energy of an isolated system remains constant. Energy can be changed from one form to another, but it can't be created or destroyed).

In this book, I have suggested that the properties of awareness and will do not derive from energy but are the fundamental ground of "zero energy," for which I like to use the term *peace*. To reiterate, for will to be free, it cannot be a consequence of *force* but in some way must relate to an absence of enforcement. After all, if you were always compelled to make all your choices by your genes and environment, then you would not have free will.

LIGHT AND THE MIND-MATTER CONTINUUM

Light has a key role in the space-time continuum as the bridge between union and separation. We have seen that, from the perspective of light or from within a light-like reference frame, there is no separation between the two ends of a light ray, either in space or in time. How about what I have termed the "mind-matter continuum"? Does light have an important role to play there too? Could it be the same one?

Consider yourself as an example. If someone looks at you, they see a form, a material structure. This body, though, somehow seems to enable a localized interaction between mind and matter. You experience sentient perception of the signals received via your senses. The words on this page are examples of such sensory information. You also exercise freedom of choice and can therefore choose to read or not to read. So awareness and will interact with matter at locations that we call individuals or people.

Awareness and will are fundamentally each of the same nature in all individuals. All that differs is the *content* of that awareness and the *choices* that are made using will.

It bears repeating that awareness and will take on these locations through association with matter. Matter and, in particular, *mass* are essential for defining location.

At this point, I would recommend taking another look at the quote from Bohm on pages 73–74.[42] If matter is *frozen light,* as Bohm suggested, then we could see the process of the *freezing* of light as the one in which separation and locations in space and time occur. Space and time themselves, when seen as separation between locations or "moments," would also be the result of that process.

Quantum physics implies that unless consciousness is involved through the process of observation, then matter is "merely" potentiality. This process of observation implies a temporal causal involvement of consciousness. Time and causation imply a distinction between subject and object and between cause and effect, and so separation is also implied by that distinction, as subject and object are separate. Once again, it may be worth clarifying that quantum theory suggests that matter is simply possibilities that are actualized through the existence of a conscious observer such as you or me. The fact that it is actualized means it is not merely an illusion, as anyone who has stubbed a toe will testify! Matter is real. But it is only real because we perpetuate an existence of separation and therefore of time and space.

The interaction between mind and matter, for example, through the action of your free choice to influence the electrical activity in your brain to tell your hand to turn the page of this book, seems therefore less mysterious. If matter is made real by sentient observers, then it is not surprising that they can also influence its activity. The theoretical physicist and quantum mechanics pioneer Werner Heisenberg wrote:

> The atoms or the elementary particles . . . form a world of potentialities or possibilities rather than one of things or facts.[56]

SUBJECT AND OBJECT

The questions raised by quantum physics encourage us to reconsider the issue of subjectivity and objectivity. The empirical method is usually seen as a way of expunging our subjectivity from experiment and observation, but for many scientists, it is the process of observation itself that is responsible for allowing us to see the world without directly encountering quantum "weirdness." In other words, according to quantum theory, the electron can behave as though it is in two places at once until we observe it. Only upon observing it is it "in one place," Schrödinger, as we have seen (see pages 2, 31, and 41), pointed out the limitations inherent in sticking exclusively to the empirical method of analysis, which is a means of studying the world as though we as sentient beings are not present in it.[2] However, the empirical method when applied to quantum mechanics puts us right there at the scene of the crime, so to speak, as indispensable, active parts of reality. As I have said, it could be argued that at a fundamental level the nature of mind is singular in as much as we all share the state of being sentient or *aware*. What differs between us is the *content* of that awareness. We all harness the power of freedom of will, but what differs between us are the choices that we make.

I have pointed out that what we call the present or the "now" is not something that emerges out of the known laws of physics (which are themselves seemingly indifferent to such a concept). The "now" is our subjective awareness. We have seen that Schrödinger considered that time itself is a derivative of mind.[2] Mind, therefore, in this worldview cannot be made in time or ended in time. Mind would have to be truly eternal and able to exist both within and without time. This would imply that as sentient beings, we have no beginning or end. The physical universe we see around us, however, would seem to have had a beginning at the big bang out of the no'thing'ness from which the initial singularity gave rise to space, time, and matter. If we had no beginning, then our roots can truly be traced right back to that original no'thing'ness.

Some people might consider this to be a contradiction in terms. How can "we" have existed in nothingness? Surely, if we were there, then that means there was "something" there, right?

Well, we need to define what we mean by "something." In our normal experience, we subjectively experience the world, and the objects of our perception—books, tables, chairs, and so on—are seen as the *things* we experience. In other words, "somethingness" implies space, time, and matter. Matter, space, and time are all properties that arise with the physical universe. We can't conceive of them existing without a physical universe, but is it possible to consider for the sake of argument the possibility that although these objects of our perception require a physical universe in which to exist, the ground state of that capacity to perceive, that is, sentience, does not require these objects to be its contents? In other words, if you consider for a moment that the perception of the objects around you exists in your mind, then perhaps you might agree that perception would still be possible for you without objects to perceive. Taking this a step further, would you allow for the possibility that perception or awareness might exist without anything physical at all?

Could it be that the traditional postindustrial "Western" view in which objectivity is seen as absolute and subjectivity is seen as relative is ill-informed?

Perhaps the objective world originates as a result of subjectivity or sentience taking a relative perspective.

This is, after all, an inference that may be reasonably taken from quantum physics that has led some leading scientists, such as Andrei Linde, to say that the physical universe might exist *because of* consciousness.

I will take the liberty of repeating Schrödinger once more.

> Dear reader, recall the bright, joyful eyes with which your child beams upon you when you bring him a new toy, and then let the physicist tell you that in reality nothing emerges from these eyes; in reality their only objectively detectable function is, continually to be hit by and to receive light quanta. In reality, a strange reality! Something seems to be missing in it.[2]

A WORLD BEYOND TIME AND SPACE

Many descriptions of near-death experiences refer to a light that people have seen. Descriptions of this light involve a sense of unconditional love and also a clear knowledge that seems to be all-encompassing. In veridical near-death experiences, people experience a perception from a viewpoint outside of their body and make physical observations during this episode that can be objectively verified.[46,47] This has been interpreted as evidence that the mind need not rely on the body or the brain as a basis for its existence.[45]

Based on the evidence that the NDE may relate to an actual experience of the mind from beyond the physical body, it is interesting to consider the ways in which the light that many people experience during NDEs is described. Is it possible that these people are in some way getting a glimpse of the connection between the material temporal universe and the state of union, of no'thing'ness, from which the universe arose at the big bang? It is perhaps worth repeating here that since the source state would be beyond time, it would have no beginning or end. I should clarify here that when I refer to the source state, I am not referring to the space-time singularity at the start of the big bang but, rather, I am referring to the state that provides the potential for this singularity to be. In other words, the singularity may be seen as an interface between the state of no'thing'ness and the physical universe.

I have made a case that light is the interface between the states of union and separation. Based on Einstein's equations we can see that from a perspective within light itself there is no separation between source and destination of the light ray either in space or in time.

I once gave a presentation titled "The Light that Shone in the Darkness" at a Turin Shroud symposium at Nicolaus Copernicus University in Toruń, Poland (the full paper is available at the Light of the Shroud website). In the lecture, I spoke about my ideas regarding free will and the clues it provides about the relationship between mind and matter. I also summarized what Bohm said about rays of light demonstrating the underlying unity behind points that appear separate in space and time. I did this in reference to the light that appears to have emanated from the body of the man on the Shroud and speculated that this could have been related to how human beings can transcend separation by "loving neighbor as self." During the question period afterward, a student pointed out something that he thought to be a contradiction in what I had said. His point was that as no time "passes" from the perspective of light, then it follows that if a man "becomes light," he has no time to make choices and so free will ceases to exist. He thought my thesis was that free will is itself the acme of human potential.

I found his comment helpful as it demonstrated a vital point that I had not clarified in detail in my presentation. Free will is the creator of time, and all active decisions occur within time and therefore within the state of separation of points (see chapter 6). The whole point about *union*, and therefore *love,* is that it relates to a state of existence beyond time and therefore, in many instances, explicit choices are not relevant.

The questioner appeared a little taken aback when I immediately agreed with him that there are no "choices" to be made in a state of existence beyond time and space, and I replied with a "down-to-earth" example of this (see the box "Time and Empathy").

Time and Empathy

Imagine that you are walking along the street and you see a stranger lose his footing and start to fall. If you have a sense of empathy and know that a fall might cause him injury and suffering, your first reaction might instinctively be to reach out to him and help. This could happen before you have *time to think* about what you are doing! As such, it could be that although you have free choice, you didn't use it in that moment. Free choice is defined as being free if you "could have done otherwise," yet in this case, you were compelled to help him because you cared. Caring was not something you *chose* to do but was instead a natural, unplanned expression of who you are.

I would argue that the fact that people are capable of unplanned responsive acts of compassion is evidence of their connection to the source state of all existence beyond space and time.

One of the themes of this book is an enquiry into the nature of mind and how this relates to our material existence. Quantum physics has, to some extent, approached the same subject matter *in reverse*. Through theoretical and empirical research in physics, many physicists have been led to the conclusion that sentient observers have a fundamental role in the foundations of existence.[2,3,30,35]

Some quantum physicists, notably including physics professor John Archibald Wheeler, have said that everything ultimately resolves to information and that the quantum description of reality is a binary series of *yes* or *no* answers to questions.[57,58] He argued that, ultimately, all physical reality is simply composed of this information. This is often known as the "it from bit" idea, where "it" refers to physical reality and "bit" is a computer term for information. I have already expressed my opinion about the nature of information (see the "Information and Knowledge" section in chapter 6, page 89). To say that the material universe is composed of information is to underline the point that the material universe might be a derivative of consciousness. It bears repeating that despite the assertions of many advocates of artificial intelligence, consciousness is *not* made of information. It is important not to confuse information with *one who is informed by* information. If I tell you that there was a red sunset here tonight, then you are informed about what I have seen, but no list of facts or data can create a mind *to be aware* of that data!

Some people like to think that the universe is some kind of computer that processes information. However, perhaps it might be worth repeating that the logician Kurt Gödel, from whose work much of computer science was originally derived, demonstrated through his incompleteness theorem that much of what is considered mathematical truth by the consensus of mathematicians is not computable as part of any algorithms but can only be seen through what he referred to as intuition.[23] Intuition is the prerogative of sentient minds and not computers.

As we have seen, it is not just intuition that is the prerogative of sentience but also any and all thinking, perception, and experience. Remember, Schrödinger himself argued that time itself is a product of consciousness, as the time in the equations of physics has no "now" and no "flow." If time is a product of mind, then mind cannot be a product of time, and therefore sentient being has no beginning and no end. If sentient being has no beginning, then it clearly cannot be created or manufactured.

As I have noted in earlier chapters, there is empirical evidence from near-death experiences that demonstrates that we do not "end" at death. In this book, I have also provided evidence that we have no beginning. Physical universes may begin our separation from each other but do not begin our existence.

Consider white light shining through a prism and the resulting rainbow spectrum illuminating a screen. The prism makes the different colors visible, but the light was there before it reached the prism. That is the marvel of each and every human being—that their existence is of such significance that it extends even beyond the entire physical universe.

Yet just watch how readily people will succumb to the belief in mind uploading. See how the deadly faith in the idea that computers are sentient and equivalent to humanity leads to our eventual replacement by them on this planet. Or don't. There may still be time to avoid this catastrophe and for humanity to survive if we learn to understand the difference between us and machines. William Blake wrote, "Hold infinity in the palm of your hand." You do. If we realize that enough, if our lives are informed by the realization that every human being is a personification of eternity, if we realize that we are to *forever* as machines are to *never*, then perhaps we will not let them replace us, and life may prevail over that which is less than dead.

We have seen that consciousness could have a primary role in making even physical existence real. This is, of course, completely consistent with the evidence from quantum mechanics.

Many theoretical physicists describe a sense of awe inspired by the fact that the universe seems so amenable to description by abstract mathematical laws, but if consciousness is the ground of existence of the physical universe, then it is perhaps not surprising that this should be so. Einstein said:

> The eternal mystery of the world is its comprehensibility. . . . The fact that it is comprehensible is a miracle.[59]

SOMETHING FROM NOTHING

Cosmologists such as Stephen Hawking and Lawrence Krauss have made the point that in a sense the physical universe is the "ultimate free lunch."[31,60] What they mean by this is that there is evidence that the universe emerged from nothing, as the total energy of the universe, including the gravitational energy, cancels out to zero. One of the most fundamental laws of nature is the law of conservation of energy (the first law of thermodynamics), and this would be consistent with this finding in that the universe began from a state of zero energy, and it still has an overall summary total of zero energy. It is often suggested that the universe could have come into existence as a result of a spontaneous quantum fluctuation, but it bears repeating that for a quantum fluctuation to occur requires the presence of an observer to collapse the wave equation.[3,30]

According to the Schrödinger wave equation, the rate of change of a zero-energy universe would be also zero, so the equation itself relies on the presence of an observer with mass.[3,30] I would interpret this as implying an observer who is an amalgam of mind and matter just as a human being is. I am not of course suggesting that there were biological human beings at the start of the universe. One could perhaps only speculate about the form that life could have taken in the very early universe, but why need we restrict ourselves to complex carbon-based life-forms? Linde also suggests that we take seriously the possibility that consciousness can exist even without matter:

> Is it not possible that consciousness, like space-time, has its own intrinsic degrees of freedom, and that neglecting these will lead to a description of the universe that is fundamentally incomplete? What if our perceptions are as real (or maybe, in a certain sense, are even more real) than material objects? What if my red, my blue, my pain, are really existing objects, not merely reflections of the really exist-ing material world? Is it possible to introduce a "space of elements of consciousness," and investigate a possibility that consciousness may exist by itself, even in the absence of matter, just like gravitational waves, excitations of space, may exist in the absence of protons and electrons? Will it not turn out, with the further development of sci-ence, that the study of the universe and the study of consciousness will be inseparably linked, and that ultimate progress in the one will be impossible without progress in the other? After the development of a unified geometrical description of the weak, strong, electro-magnetic, and gravitational interactions, will the next important step not be the development of a unified approach to our entire world, including the world of consciousness?

All of these questions might seem somewhat naive, but it becomes increasingly difficult to investigate quantum cosmology without making an attempt to answer them. Few years ago it seemed equally naive to ask why there are so many different things in the universe, why nobody has ever seen parallel lines intersect, why the universe is almost homogeneous and looks approximately the same at differ-ent locations, why space-time is four-dimensional, and so on. Now, when inflationary cosmology provided a possible answer to these questions, one can only be surprised that prior to the 1980's, it was sometimes taken to be bad form even to discuss them.

It would probably be best then not to repeat old mistakes, but instead to forthrightly acknowledge that the problem of conscious-ness and the related problem of human life and death are not only unsolved, but at a fundamental level they are virtually completely unexamined. It is tempting to seek connections and analogies of some kind, even if they are shallow and superficial ones at first, in studying one more great problem—that of the birth, life, and death of the universe. It may conceivably become clear at some future time that these two problems are not so disparate as they might seem.[3]

In his 1932 book *The Mysterious Universe,* Sir James Jeans wrote:

The universe can be best pictured . . . as consisting of pure thought. . . . If the universe is a universe of thought, then its creation must have been an act of thought.[61]

DETERMINISM, RANDOMNESS, AND FREE WILL

Newtonian mechanics is usually considered to be deterministic in nature, and this encouraged many people in the eighteenth and nine-teenth centuries in particular to believe in a "clockwork," mechanical universe. It has been pointed out by biologist Rupert Sheldrake that this mechanistic view is actually highly anthropomorphic, as the notion of a machine, from which it is derived, is a man-made construct.[62] It is inter-esting that this worldview coincided with the Industrial Revolution and mechanization, although, of course, Newton's discoveries in mechanics may well have facilitated many of the developments in engineering that brought this about. A distinction here should be made between deter-minism and predictability. Determinism would mean that the future is completely fixed in advance, but this wouldn't necessarily mean that it was predictable. Even in classical Newtonian physics,* small uncertain-ties or errors in the knowledge of initial conditions, for example, can

---

*The term *classical* is often used by physicists for those aspects of physics that predate the discovery of quantum physics or do not directly take account of quantum effects.

quickly escalate to a great deal of uncertainty in a prediction about how a dynamical system will develop.

Quantum mechanics, on the other hand, overturned notions of determinism, both through the uncertainty principle of Heisenberg and through the inherent apparent randomness that occurs *when a measurement is made* at the quantum level. It is interesting that this randomness only occurs when a measurement is made. However, between measurements, the Schrödinger equation evolves completely deterministically through what is known as *unitary evolution,* which is referred to as **U** in the published works of Roger Penrose. **U** is distinguished from **R,** which is a term he uses to refer to the reduction of the quantum state vector, otherwise known as the "collapse" of the wave equation. This *collapse* or *reduction of the quantum state* (that is, **R**) is what happens when an observation is made. Penrose points out that although **U** is a process that it *reversible* in time, **R** is not reversible. He points out in his book *The Emperor's New Mind* that many physicists (including himself) had previously assumed that **R** was reversible, but he demonstrates that it is not.[23] This is a crucial observation. I find it fascinating that this process that is not time-reversible is at the root of the measurement problem in quantum mechanics. If, as the eminent physicists Fritz London and Edmond Bauer and later Eugene Wigner and more recently Linde argued, consciousness is needed to collapse the wave equation, then it would follow, as **R** is implicated in the origin of the arrow of time *and* is dependent on consciousness, that *we as conscious beings are both the creators of the "now" and of time itself!*

Some scholars, such as, notably, Michio Kaku, professor of theoretical physics at the City College of New York, have argued that the indeterminacy inherent in **R** proves free will. As you may have noticed from reading this book, my personal opinion is that free will definitely does exist, but I have to say that, having listened to Kaku speak, I felt that he didn't make his case very strongly. He was explaining that because of quantum indeterminism, it is not possible even in principle to predict what you are going to do. However, lack of predictability to my mind does not, of itself, without further explanation, imply free will as a direct causative agent. It is, in principle, impossible to predict exactly when a radioactive atom will decay, for example. All we can do is write down the statistics that describe how likely we are to detect decay over a certain timescale, but not many people would believe that the radioactive atom is, itself, freely *choosing* when to decay. So an important distinction needs to be made, in my view, between a lack of predictability and causation through free choice. Notwithstanding my reservations about Kaku's argument, experiments on quantum mechanics do suggest, according to some leading theoretical physicists, that the experimenter's choice of which measurement to make needs to be regarded as free in the sense that he or she "could have chosen otherwise."[36] The photon or the electron can be found to be either a wave or a particle depending on how it is measured, and the choice of how it is measured will determine which of these it is found to be. If the choice did not determine this, then that would imply some kind of bizarre determination of the choice of the experimenter, not by chemicals or electrical signals in the brain but actually by the putative preexisting state of the electron in the experiment. In other words, the correlation between the choice of experiment and the wave/particle nature of the electron implies one of two possibilities: either the choice of how to measure the electron determines how it is found to exist, or our perception of free choice is an illusion. The latter possibility would mean that every time the electron is a wave, we are compelled to perform the experiment in such a way as to be able to detect wavelike interference to confirm it as a wave, and when it is a particle, we are subconsciously compelled to look for it in a particular place and thus discover it to be a particle![36]

It is important at this point to make a distinction between the ways in which the action of free will might be considered to influence the electron. I am not suggesting that, under normal circumstances, we determine where we will find the electron to be when we measure it but that instead we decide whether we will find it to be a wave or a particle. We decide this simply by our choice of measurement process.

If "mind over matter" is involved in this, it is through controlling the processes in the brain that lead to movements of the body in such a way as to put into practice our chosen method of measurement. Once the measurement is made, there is still an element of apparent statistical randomness involved in where we find the electron. If it has behaved as a wave, this randomness "determines" where in the pattern of interference fringes we will find it, and if it has behaved as a particle, then this "determines" which of the two slits the electron has gone through.*

Could randomness itself also have an explanation in relation to freedom of will? If it does, then perhaps we have to bear in mind the "nonlocal" nature of the interaction that might be involved in this.

THE STORM AND THE BUTTERFLY

You might already be familiar with the often-quoted example from chaos theory that a butterfly flapping its wings in the Amazon rainforest could cause a storm off the coast of Florida.[63] I would not suggest that the butterfly could *choose* to make a storm, but what if it could choose whether to flap its wings? If a decision made in free will is indeed the only type of primary causation, then it would follow that all effects are directly (or mostly indirectly) the consequence of choices. For example, I have argued that the origin of the physical universe itself could have been a consequence of the choice of sentient being to experience separation, with space, time, and matter being perhaps the external objectification of this separation. Multiple experiments have demonstrated a small but statistically highly significant effect of conscious intention on random-number generators.[64] If there is any effect at all, then one interpretation of these results might be that all ostensibly

---

*For those who would be interested in a more detailed description and analysis of the double slit experiment I recommend the excellent popular science book *Quantum Enigma* by physicists Bruce Rosenblum and Fred Kuttner (Oxford: Oxford University Press, 2011).

random events are ultimately the indirect consequence of choices made in free will. Just as the butterfly didn't choose to cause a storm, we are not generally "choosing" the position at which an electron is found or the moment of decay of a radioactive particle, but these are nevertheless ultimately the summary of all choices made by all sentient beings. If this happens in a chaotic way (like the butterfly and the weather systems), then under most circumstances, the statistics would still approach randomness for all practical purposes. However, there would ultimately be no such thing as true randomness!

The influence of free choice on the apparently random probabilistic results of measurement would be something that would have to be considered a "hidden variable" in quantum theory. It has been demonstrated empirically that the results of experiments in quantum mechanics could not be explained by any *local* hidden variable theory, but it is conceivable that the effects of free will would be nonlocal in the sense that minds might not be truly localized. Our minds arguably appear to be localized *in our heads* only as a result of our viewpoint through the physical senses. Four of the five senses rely exclusively on sense organs in the head, and this has a tendency to suggest to us that our experience occurs somehow in the head. Many people would assume, of course, that since the brain is in the head that this must be the location of our consciousness, but I would argue that the reasoning behind this assumption is faulty, and I have illustrated this by the analogy to an assumption that might be made that a song is sung by a radio. It bears repeating that the evidence from near-death experiences is also highly suggestive that the solely brain-centered view of consciousness is flawed.[45,46,47]

Experimenters on entangled particles have, as stated above, proven that if hidden variables exist, then these would have to be "nonlocal." In 2011, Colbeck and Renner published a paper in which they argued that under the assumption that experimenters have free will to decide how to conduct their measurements (such as whether to look for the electron as a wave or a particle), there cannot be a level of description of reality deeper than quantum theory. In this book I have argued that an understanding of the primacy of sentience and free will leads to a deeper understanding of reality that underlies quantum mechanics and indeed underlies all reality. This would be consistent with Colbeck and Renner's assumption. However, perhaps they do not realize that the existence of free will is itself the clue to an understanding of the universe that goes beyond that afforded by quantum mechanics. I am also not sure whether they are aware that if they are right that free will is real, then the notion of materialism must be false, as free will would mean that our choices cannot be completely determined by material causes.[65]

In 1978, Wheeler proposed a thought experiment based on *delayed choice* in which the choice of whether to look for the electron as a particle or a wave is delayed until *after the electron had already passed the two slits*. This experiment was performed and then reported in the peer-reviewed scientific journal *Science*.[66] It was found that this choice could indeed be delayed, and it was demonstrated that the *history* of whether the electron had passed as a wave through both slits at once or as a particle through only one slit could be created *after the event*. This has been described by some people as causation backward in time. Is it really backward in time though? To call it this implies a belief in a real time that flows asymmetrically according to time's arrow.

It implies a belief in the notion of an objective meaning to time's arrow independent of the conscious observer.

Penrose pointed out (in a different context) that "*consciousness is, after all, the one phenomenon we know of, according to which time needs to 'flow' at all!*"[23]

In Wheeler's delayed-choice experiment, we have an empirical clue about the nature of time.[66] Remember, Schrödinger pointed out that "*Mind is always now,*"[2] and that the "now" and therefore time itself are products of consciousness. This would be consistent with the results of the experiment inasmuch as they imply that *the "now" and the arrow of time only exist where information is available to conscious observers.*

"Superposition" states of potentiality prevail where information is not available to temporal conscious observers.

Scientific evidence suggests that the entire physical universe may well have begun from out of nothingness.[31,60] If time is a property of consciousness, then consciousness could not be created by or in time.[2]

It is clear that it doesn't make sense to talk about existence "before" time, as the word *before* implies the presence of time. However, consider the possibility of there being two states of being, one temporal and one timeless. To be timeless is to be eternal in the sense of having no beginning or end, so the timeless state exists *now* just as the temporal state does. Time in the sense of separation and causality is dependent on the existence of a timeless state of being, but no change, and therefore no time, is required for timelessness to exist.

It is interesting to compare the properties of the two states of being from the point of view of the observer. If you look around you now, you might see objects and people apparently existing in separate locations in space and with defined properties such as size and shape as well as states of apparent stillness or motion. Both stillness and motion imply time. The fact of an observation of "somethingness," of matter and space, also implies time. So time is the product of separation, and that is why it arrives in a continuum with space, which is simply the separation of points. From your point of view as you are reading this book, consciousness is experienced as though it is made up of separate moments of sentient observation from a particular viewing point, which I call "you," and you (in reference to yourself) would use the pronoun "I." This is all consistent with the apparently "classical" world emerging from the collapse of the Schrödinger wave equation as a result of consciousness existing in a separated temporal aspect.

Can we speculate about what it might be like to experience consciousness from the timeless state? We can infer some of the characteristics of this, although perhaps we cannot really appreciate them in their true nature as long as our minds are cultured in separation and division. There could be no division between subject and object from within this state of being, and there would be no things—no space, no time, no matter. So in what sense would you exist as an individual? Bohm once made an interesting observation about the word *individual*. The word literally means "undivided" in the sense that you are seen as an undivided unit or atom of consciousness, as it were. (The word *atom* also means "undivided" and was coined in ancient Greece by Democritus to convey the notion of the smallest possible [indivisible] unit of matter.) However, as Bohm pointed out, the word *individual* is usually used to denote the division of consciousness between one identity and another.[42] To be true individuals, we would have to be undivided from each other also. One could consider this in the following way.

We perceive a world of things with properties seen in terms of matter, space, and time, but we are also aware of our existence as perceivers and of the existence of other sentient beings (such as the people around us). We perceive these people and things as having discrete and defined properties from moment to moment but always perceived by us within the "now." The presence of these defined properties as parts of our experience means that we can perceive change, which we interpret as a subjective flow of time. In other words, because we perceive an apple to have a defined location, we can experience a change in that location, such as when it falls from a tree. Because we experience change, we have the notion of the arrow of time and the apparent flow of time.

However, from a perspective beyond time and space there would be no things. There would be perception, but the perceiver would be undivided and without limit. I have made a case that the fundamental ground of perception is the same for all of us, but all that differs between us are the particular patterns of memory, experience, and choices that define our differences from each other. Perhaps a clue to the other state of being is contained within the light that people perceive in near-death experiences. The light is often described by experiencers in terms that refer to its wholeness, completeness, oneness, undivided knowledge, and unconditional love. Perhaps within that light you exist as you, but that *you* is more truly individual because it is not divided. It is not divided because it is fully inclusive of *all* sentience, and therefore of all sentient beings and of all knowledge. The one thing it doesn't know is ignorance, as it can't be ignorant if it is all-knowing.

I am not merely making a semantic distinction here, but I would contend that this is a logical necessity. It is our patterns of restriction and ignorance that differentiate and divide us from this light. If our fundamental nature as sentient beings is for us all to be *one,* then qualities such as selfishness, materialism, and prejudice are what preclude us from perceiving our true original nature and thereby create the darkness that keeps us away from that light. Since ignorance is a state of "not knowing," then one who is all-knowing will not experience ignorance. I would argue that we never really "know" our ignorance, but instead our ignorance is a summary of what we do not know! Clearly, I am relating ignorance here not to lack of knowledge of facts or data but to lack of knowledge of "that which knows," for example, to a lack of empathy for other sentient beings.

Could it be that all that is known is known through that "light" within us, in whatever dregs of that light that exist within us, as the light is the "screen" of awareness itself on which all our perceptions are written? Seen in this context, the saying "the light shines in the darkness and the darkness did not understand it" would make sense, as it is only through the light that understanding exists.

If, as I have argued, the seat of awareness is beyond space and time, then the "light" I refer to would be the interface between awareness and the temporal universe. Einstein said:

A human being . . . experiences himself, his thoughts and feelings as something separated from the rest—a kind of optical illusion of his consciousness. This delusion is a kind of prison for us, restricting us to our personal desires and to affection for a few persons nearest to us. Our task must be to free ourselves from this prison by widening our circle of understanding and compassion to embrace all living creatures and the whole of nature in its beauty.[67]

Here and Now

A Time-Mind-Space-Matter Continuum

This chapter is intended to summarize and bring together some of the key points of the previous chapters and to show how they lead to some remarkable conclusions. It is based on some papers that I wrote and presented to the Scientific and Medical Network and the Society for Scientific Exploration in 2012 and 2014.

We all habitually describe where we are as "here" and "now." It is my contention that if we understand here and now, we will better understand mind and the material universe as well as the connection between the two.

Schrödinger pointed out that "Mind is always *now*" (my emphasis).[2] He derived from this the argument that mind is eternal, and he also elaborated on this to argue that at the fundamental level all mind is one.

Penrose has suggested that we will understand consciousness better once we have arrived at a unification between quantum theory and relativity.[49]

Lee Smolin has pointed out that the laws of physics as currently understood don't imply in themselves either the existence of the "now" or what we experience as the flow of time.[43,68]

He also cautioned against the logical flaws inherent in seeing space and time as a given—as simply the "when and where" of substance and events.

I have pointed out that in considering the relationship between these four elements (mind, matter, space, and time), we would do well to acknowledge an "elephant in the room."

If there is such a thing as freedom of will, then this could perhaps be a clue to the nature and properties of all four elements.

It is interesting to note that clues regarding twentieth-century breakthroughs in relativity and quantum mechanics were already present in the ideas of great scientists centuries earlier. Newton considered light to be composed of quantized particles. Galileo's experiment in Pisa implied the equivalence of gravitational and inertial mass, which was later a basis for Einstein's theory of general relativity.

What might be the hidden clues in current physics regarding the next great breakthrough?

There have been many triumphs in natural science in the last few centuries, the most recent being the discoveries of relativity and quantum mechanics early in the twentieth century. Since then there have arguably been no major breakthroughs in underlying principles, but instead there has just been some success of the "particle spotters" in finding the predicted particles of the "standard model," culminating in the Higgs boson. More recently, there has been the groundbreaking discovery of gravitational waves, but these had already been predicted a hundred years earlier by Einstein's equations of general relativity and therefore were a breakthrough in measurement rather than of understanding. Nobody has been able to find a way to reconcile quantum mechanics with general relativity. There is no generally accepted clear solution to the measurement problem of quantum mechanics or to what philosopher of mind David Chalmers calls the "hard problem" of consciousness.

We still have no clear consensus understanding of the hard problems of time. Did it have an origin? If there was an origin of time, then what caused it? Why and how is there a present such as we refer to when we say "now"? How and why do we seem to experience a flow of time? What is the explanation of the hugely ordered beginning of the universe, which appears indispensable to the origin of the arrow of time?

If consciousness is (as some of the pioneers of quantum theory believed) indeed fundamental to the nature of reality, then could we find clues to the answers to these questions in the inward and outward aspects of consciousness, namely awareness and free will?

Let us again consider these phenomena of awareness and free will and how they might relate to quantum mechanics and Einstein's relativity. As Arthur Eddington said:

> The physical world is entirely abstract and without "actuality" apart from its linkage to consciousness.[28]

CONSCIOUSNESS AND THE COLLAPSE OF THE WAVE EQUATION

The Schrödinger wave equation tells us that physical reality consists of superpositions of different properties that exist together. For example, in the case of the decay of a radioactive particle, the particle can have both decayed and not decayed at the same time. As macroscopic (large scale) objects are composed of microscopic parts, which are described by the equation, there is an implication that these objects should also exist in superposition states. Although this is what is predicted by the wave equation, Schrödinger himself found this to be preposterous.

....

Schrödinger's Cat

Schrödinger illustrated his misgivings with a famous gedankenexperiment, or "thought experiment," in which the decay of a radioactive particle would be detected by a Geiger counter that would then trigger the release of a capsule containing hydrogen cyanide gas into an opaque box containing a cat.[69] If the particle decayed, then the cat would be killed, and if it did not decay, then the cat would remain alive. However, since the particle existed in a superposition of states of having both decayed and not decayed simultaneously, it would follow that the decay would be both detected and not detected. The poison would then be both released and not released, and the cat would then be both killed and still alive simultaneously. Schrödinger was poking fun at his own equation as he felt that its predictions could not be fully consistent with reality.

....

However, the equation has been shown to be verified very precisely over and over again by empirical observations. Observation of a quantum state appears to collapse the wave function, which through this process is somehow transformed into one that describes the probabilities of the various outcomes of observation. However, there is still the unresolved issue of what constitutes an observation or a *measurement*. It is interesting that in the early days of the elucidation of quantum mechanics and relativity, many of the pioneers of these fields did believe that consciousness was fundamental and that consciousness was required for the wave equation to collapse.

In a 1931 interview for the London *Observer* newspaper, Schrödinger stated:

Consciousness cannot be accounted for in physical terms. For consciousness is absolutely fundamental. It cannot be accounted for in terms of anything else.[70]

Sir James Jeans, in his 1932 book *The Mysterious Universe,* wrote:

> Mind no longer appears as an accidental intruder into the realm of
> matter; we are beginning to suspect that we ought rather to hail it
> as the creator and governor of the realm of matter.[61]

Much more recently, Linde wrote:

> Without introducing an observer, we have a dead universe, which
> does not evolve in time. Does this mean that an observer is simulta-
> neously a creator?[3]

Some of the pioneers of quantum mechanics, such as John von
Neumann and Eugene Wigner, pointed out that the objects that form
the physical measurement apparatus, being themselves composed of
quantum mechanical parts, could not in and of themselves collapse the
wave equation and therefore that a conscious observer had to be at the
end of the "von Neumann chain" of observation. This was because conscious observers never directly find the objects of their observation to
be in a superposition of states. Instead, they always perceive them to
have collapsed to one or another of the possibilities when the state is
directly measured. The wave equation evolves in time in a completely
deterministic manner that is technically called "unitary evolution."
This describes the simultaneous evolution of all the various possibilities
together in superposition, and it is only through observation that it is
reduced to one actual outcome. This can be seen as the *actualization of
potentiality through observation.*

This leads us to some interesting observations regarding time. We experience time always in the present. To quote Schrödinger again, "Mind
is always *now.*" One could argue that the past and the future only exist
within the present. The past exists within the present as *actuality*—
the current state of the way things are, which has arisen through past

events. The future exists within the present as *potentiality.* So conscious-ness makes the potential actual via the collapse of the wave equation. That would be consistent with the statement that "Mind is always *now*" if the present is where potentiality and actuality meet. If awareness is the *inward* aspect of mind, then one could see free will as the *outward* aspect, judged in terms of action through volition.

Schrödinger, as we have seen, argued that because "Mind is always *now*," time would seem to be a derivative of mind, and as such, mind could not begin or end within time and must therefore be eternal. He wrote:

> So with all due acknowledgement to the fact that physical theory is at all times relative, in that it depends on certain basic assumptions, we may, or so I believe, assert that physical theory in its present stage strongly suggests the indestructibility of Mind by Time.[2]

He also pointed out that although we commonly consider mind to exist in the plural, the fundamental nature of it would seem to be singular and unified.[2]

To illustrate this, we can infer that a state of "separateness" had a beginning when the physical universe emerged from nothingness at the big bang. If mind is eternal but separateness had a beginning, then all mind must be one within the timeless singular state of no'thing'ness and all potential, from which the physical universe emerged.

So there are three different ways in which the possible becomes the actual—in which what "*could be*" becomes what "*is*."

One is through the quantum effect, whereby conscious observation col-lapses the wave equation to materialize an actual reality out of the "sea" of quantum potential.

The second is through the action of free will, whereby one option is chosen out of several potential options.

The third is the one we perceive as time, a present in which out of many possible futures one becomes manifest.

My suggestion is simply that these three are three facets of a single reality. They are clues to the nature of the physical universe and also to the nature and potential of sentient beings. This would have profound implications for the significance of us all as it places all sentient beings, including us, at the heart of the foundation of reality itself!

According to Einstein's equations, light and all else that has no "rest mass" exists at the interface between time and eternity. This is because at the speed of light, the separation of points that defines space and time ceases to exist as origin and destination become one and the same.[42] If mind has no intrinsic mass, then perhaps this also is a clue to our true potential scope of existence.

TIME AND CAUSALITY

I have argued in this book that the exercising of free will implies that sentient being and sentient being alone has the power of primary causation. All other causes would be secondary derivatives of what are known as initial conditions (see box "Causation").

....

Causation

To give a simple example of primary causation, if you choose to knock over one domino that is aligned in a chain with numerous other dominoes, then your decision is a primary cause for why the first domino fell. The fall of all the other dominoes could be seen as a chain of secondary causation related both to the fall of the first domino and the juxtaposition of the whole set of dominoes in relation to each other.

....

In relativity, time and causality are intimately related. A location in space and time such as what you might call "here and now" as you are reading these words is known in relativity as an *event*, and two events are said to be *causally connected* if one could move from the first to the second without exceeding the speed of light. However, the notion of causality is generally assumed to include the arrow of time in that an *earlier* event can have a causal influence on a *later* event, but the converse is not true. This is tied in with our subjective experience of what is often termed the *flow* of time, which implies that the *past* has *passed* and is already determined and *actual,* whereas the future is not yet determined and exists in the form of *potentiality* such that present events can determine future ones. The delayed-choice experiment demonstrates that at the quantum level there is no fixed arrow of time.[66] Here again, the sentient observer is implicated in making time "directional" by the collapse of the wave equation.

Gödel made a very significant discovery about time that is often overlooked. He showed that Einstein's equations predicted that it was possible under certain circumstances to create *closed timelike curves,* which basically imply that one could travel to one's own past, creating a circular looped path through space-time. As Palle Yourgrau, professor of philosophy at Brandeis University, has pointed out, this would not be a recipe for time travel but would actually mean that time did not exist.[71] If the past is somewhere you can visit, then it has not *passed* as it is still present, and all sorts of logical paradoxes would follow. This prompted Stephen Hawking to propose his *chronology protection conjecture,* which suggests that there must be some as yet undiscovered law of nature that prevents travel into the past.[72] Some have speculated that once quantum theory and relativity have been reconciled to provide a coherent theory of *quantum gravity,* this will clarify why it is not possible for an object to travel to its own past.

If I am right that awareness and will are fundamentally the progenitors of all existence, including space and time, then perhaps quantum gravity will only be understood once they are understood.

QUANTUM WEIRDNESS, EINSTEINIAN WEIRDNESS, AND COMMON SENSE

It could be argued that the reason why we find notions such as non-locality and relativity to seem weird or counterintuitive is that we interpret our sensory perceptions as implying that there is a world of substance "out there" in the world that is fundamentally real in and of itself. It is easy to forget that these sensory perceptions are experienced and realized within our minds, which, themselves, are *not* substantial objects but without which no experience of substance exists. As Schrödinger observed, the empirical view of the world has been defined by excluding us, as sentient observers, from its contents, and therefore, we do not find our minds to be within its confines.[2] Rather, it is the other way around, and the empirical world only exists as such because of conscious observers. Eddington reminds us:

> It is difficult for the matter-of-fact physicist to accept the view that the substratum of everything is of mental character. But no one can deny that mind is the first and most direct thing in our experience, and all else is remote inference.[28]

Awareness, free will, empathy, and the "now" are part of our everyday common-sense view of the world. If these things are real *as we observe them to be,* then the apparent weirdness of quantum mechanics might simply be a product of the discrepancy between our artificial, classical, commonsense notions of the real world out there and the fact that at the fundamental level, it is sentience itself that underpins all existence. I would argue that even the strangeness of the apparent quantum paradoxes is eclipsed by the strangeness of the bizarre and often-expressed notion that people became self-aware because they had words for self-awareness. People actually write books making statements like this, saying that sentience evolved after language and is created by language. They believe that we became aware because we had words like *I* and *me*. If they are correct, then their books ought to be sentient beings as they contain such words!

As language is a means of communication between people, it has a way of enabling the expression of shared ideas through its common usage. Take the word *nothing*, for example. Literally, this implies the absence of things or objects. However, this relates to the absence of matter and needn't imply the absence of possibilities or potentiality.

The whole physical universe appears to have emerged from out of nothingness at the big bang,[60] so it would seem that rather than being empty of possibilities, nothingness might actually be the true cradle of potentiality. I have also argued that our sentience or awareness as conscious beings is not a projection of our brains and not a product of any object or thing. My view is that awareness and freedom of will are natural properties of no'thing'ness.

Let's take another look at the phenomenon of nothingness, beginning with the number zero.

ZERO AND NOTHING

Many people are surprised by the fact that before the year 1202, there was no known reference to the number zero in Europe. It was introduced to Europe by Fibonacci (Leonardo of Pisa), who had grown up in North Africa and received his education there in what is now Algeria. African scholars had been familiar with the number zero (in Arabic, known as *sifr*) for centuries, and so he learned about it from them. He brought this knowledge back with him to Western Europe. This knowledge had been brought to Africa by Arabs from India, where it appears to have first been discovered around the fifth century BCE. At that time, Pingala and his contemporary Indian scholars referred to it as *śūnya* (pronounced "shoonya"), which translates as "empty" or "void," and Indian mathematics of the time included this number.

It is interesting to consider the views of Lawrence Krauss regarding "something" and "nothing." He makes a strong case that the physical universe emerged from nothingness.[60] Edwin Hubble discovered in the 1920s that the physical universe is expanding uniformly in all directions. There is no single center of this expansion, although one could view all points as being central. A Belgian priest named Georges Lemaître suggested the big bang theory. Lemaître studied under Eddington. Both Eddington and Einstein were initially skeptical about the big bang theory, but both were very quick to accept it after examining Hubble's evidence pointing to the expansion of the universe. Krauss points out that quantum theory allows for the appearance of "something" from empty space. Empty space is referred to as nothingness by Krauss, but where did space come from, and why does quantum mechanics apply to it?

I have explored these questions in more detail elsewhere in this book. Einstein discovered that gravity curves space. To explain what the curvature of space is, it is helpful to bear in mind that light appears to move in straight lines. Well, it actually moves in the straightest *available* lines. This can be visualized as a plane flying in the shortest possible distance at a constant height between two locations on the Earth's surface. Such a line is known as a *geodesic*. It is because gravity simply describes the straightest paths for matter and light through curved space-time that physicists point out that gravity is not really a force. It only looks like a force to us because we can't see the curvature of space-time. As I noted in chapter 5, Einstein predicted something called *gravitational lensing*, whereby the curvature of space by a massive object will cause light to be bent *around* the object, which therefore acts like a lens. This was empirically confirmed by Eddington during a solar eclipse in 1919.

Overall, though, evidence suggests that the universe, when considered as an average of all observable space, is not curved but is *flat*. For a flat universe, the total energy, including gravitational negative energy, is zero. This means that the universe could have appeared from nothingness without violating the law of conservation of energy.[60] The total energy began at zero, and it is still zero. The whole thing is a construction of nothingness.

However, this presupposes the existence of the space in which the fluctuations occurred, and it also presupposes the preexistence of quantum mechanical fluctuations.

Is this a reasonable supposition, or is there a more rational explanation? Wheeler says:

> The Universe had to have a way to come into being out of nothingness. . . . When we say "out of nothingness" we do not mean out of the vacuum of physics. The vacuum of physics is loaded with geometrical structure and vacuum fluctuations and virtual pairs of particles. The Universe is already in existence when we have such a vacuum. No, when we speak of nothingness we mean nothingness: neither structure, nor law, nor plan. . . . For producing everything out of nothing one principle is enough.[73]

FROM ETERNITY TO NOW

To quote Schrödinger:

> Eternally and always there is only now, one and the same now; the present is the only thing that has no end.[74]

How should we characterize a state of existence beyond space, time, and matter? Some might ask whether it even means anything to ponder the existence of such a state as they might consider that all reality is defined in terms of these three parameters!

I would argue that we *know* of space, time, and matter *because* we are fundamentally beyond all three.

As we have seen, the laws of physics don't in and of themselves imply such a concept as the flow of time.[43] Nor do they account for the existence of a present in the sense of past, present, and future.

As I have mentioned previously, Schrödinger[2] argued that time is a product of consciousness. Also, Penrose pointed out that the flow of time is only relevant within consciousness.[23]

So could it be that the "now," the origin of time, the flow of time, and the arrow of time are all derivatives of consciousness?

Let's consider these one by one.

1. The "Now"

It is often suggested that classical physics applies to the past and quantum mechanics to the future, as quantum mechanics relates to a world of potentialities and classical physics describes a world of actualities. If that is so, then a case could be made that the observer then has a crucial role in defining the present. Not merely in the content of the present but also in the fact that such a thing as the present exists at all!

Empirical evidence from such sources as Wheeler's delayed-choice experiment encourages us to consider more deeply what is meant by any suggested distinction between the present and the past.[66] The experiment shows that the *history* of a particle can be influenced by a choice of measurement made *after the fact*. Whether a photon had traveled on a single path or on two available paths can be influenced by a choice of measurement that is made *after* the photon has "made its decision" about whether to behave as a wave or a particle (i.e., whether to come via both paths or only one). (Please note that I am not suggesting that the photon is literally making a decision, but it is difficult to express this apparent paradox in words, as language is based on our experience of causality and distinctions between past and future.

It is we as sentient beings who make decisions.) It seems to me that an inescapable conclusion from all this is that *the notion of the "now" being a product of consciousness is confirmed by experimentation.* People talk as though the observation has an influence backward in time, but that presupposes that the passage of time is objective for the "real world out there," independent of the existence of consciousness or the observer. This experiment would seem to rule out this hypothesis and suggests instead that the history of the path of the photon only comes into existence once it is *observable.* To quote Wheeler regarding the significance of this:

> Thus one decides whether "the photon will have come by one route, or by both routes" after it has "already done its travel."[73]

Wheeler considered that the universe is written in information, and he coined the term "it from bit." However, as I have said, I would contend that considerations regarding the nature of time and consciousness suggest that there must be a deeper level of explanation than "information." If, at a fundamental level, there is no information without sentient awareness, then awareness must be more fundamental than information.

A report on an empirical experiment demonstrating Wheeler's delayed-choice idea was published in February 2007.[66] Wheeler died shortly afterward in 2008 at the age of ninety-six.

2. The Origin of Time

Schrödinger himself realized that if the "now" is the product of the observer and therefore a product of awareness itself, then the mind cannot begin or end in time.[2] However, evidence does suggest that the physical universe, including time, space, and matter, did have a beginning around 13.8 billion years ago. This would imply that mind does not depend on the existence of space, time, or matter for its own existence. What would the properties of mind be without space, time, and matter?

Perhaps there is a clue to this in the nature of *light*. According to relativity, from the reference frame of a light beam, there is no separation in space or in time between the origin and the destination of the beam. For this reason, the path of the light beam through space and time is known as the null path. Bohm suggested that matter is a derivative of light.[42]

So, according to Einstein's equations, light and all else that has no rest mass exists at the interface between time and eternity. This is because at the speed of light the separation of points that generates space and time ceases to exist, as origin and destination become one and the same.

There is some evidence that in the periods soon after the origin of the universe and also in the far future at the "heat death" of the universe, the universe was and will be dominated by photons rather than matter.[40] Light does not have "clocks" because there is no passage of time from the perspective of the photon, so time begins as mass begins. What causes light to be frozen in forms that have properties of mass? Another way of posing the question is: *What causes a temporal physical universe to begin?*

Some people postulate a creator being who set the ball rolling, as it were, by an act of creation. However, this displaces the question back one step as it doesn't account for where the creator being came from or why he, she, or it would do this. Also, the creator being would by necessity have to become temporal itself if it is acting "in time" by creating as there would then be a before-and-after distinction for it (i.e., before and after the moment of creation.) If the creator being were identified with the very state of timelessness from which the universe emerged, then it would seem that that state of timelessness would then have to end once the creator being became temporal. This is, of course, a contradiction in terms as it is in the nature of timelessness that it cannot have beginning or end. People may well posit a creator being who "inhabits" timelessness rather than one who "is" timeless, but even if that were the case, then the creator would have to be a temporal being, and surely this is another contradiction in terms!

Others, such as Krauss and Hawking, have suggested that the emergence of something from nothing and of time from timelessness is a natural consequence of the laws of physics and results from quantum fluctuations in empty space. My contention would be that Wheeler was right when he pointed out that nothingness implies that even empty space could not have been present (see page 145).[73] Space is, after all, merely the separation of points. Points in space exist as locations in space because of their separation one from the other, which arises as a result of light being frozen as per Bohm's description.[42]

So, in considering the nature of consciousness beyond space, time, and matter it would seem that it would make sense to consider the possibility that *we* might all be *one* within that state. Light would be the physical correlate of the interface between the timeless singularity state of no'thing'ness from which the universe emerged and the temporal physical universe as we know it now. It is interesting that if Schrödinger was right about mind existing beyond time as well as within time, then he would also have been right that mind cannot begin or end in time. One implication of this is that physical death is not the end of us as sentient beings. With this in mind, it is interesting that many people in their reports of near-death experiences refer to an experience of an incredibly bright light that they characterize as containing all knowledge, and they describe a clear sense of the light emanating profound unconditional love. How would we define *love*? Well, in this context it seems to imply a unity between sentient beings. It is generally considered that empathy, which is the capacity to feel the joy and suffering of others, is an expression of it. So, what if that were taken to its full extent? Well, it would mean that all sentient beings would be one and experience the *"all"* instantly from all viewpoints together. If that were the case, then there would be no need for separation or for space, time, or matter.

3. The Flow of Time

Rudolph Carnap, the Viennese philosopher, recounts a fascinating dialogue he had with Einstein about the nature of time:

> Once Einstein said that the problem of the Now worried him seriously. He explained that the experience of the Now means something special for man, something essentially different from the past and the future, but that this important difference does not and cannot occur within physics. That this experience cannot be grasped by science seemed to him a matter of painful but inevitable resignation.

Carnap replied in the following way:

> I remarked that all that occurs objectively can be described in science; on the one hand the temporal sequence of events is described in physics; and on the other hand, the peculiarities of man's experiences with respect to time, including his different attitude towards past, present, and future, can be described and (in principle) explained in psychology.

Carnap reports that Einstein didn't accept this assertion:

> But Einstein thought that these scientific descriptions cannot possibly satisfy our human needs; that there is something essential about the Now which is just outside the realm of science.[75]

I would agree with Einstein that psychology cannot, even in principle, explain our sense of the "now" or of the flow of time, as psychology deals only with the contents of our perception and therefore cannot account for the fact that we have the *capacity* to perceive at all. However, whether something is outside the realm of science might depend on how widely we define the realm of science. My suggestion in this book has been that science (a word based on the Latin word *scientia,* meaning "knowledge") might become broader in its scope if more of its proponents were to consider the possibility that consciousness might be fundamental to existence itself. Once that occurs, then perhaps more scientists will acknowledge, as did Schrödinger, that the "now" might be a property of pure awareness and that time is itself an emergent product of consciousness!

Smolin, in his 2013 book *Time Reborn,* makes an interesting comment about the Einstein-Carnap dialogue:

> Einstein and Carnap agree about one thing: that experiencing nature as a series of present moments is not part of the physicists' conception of nature. The future of physics—and, one might add, the physics of the future—comes down to a simple choice. Do we agree with Carnap that the present moment has no place in science, or do we follow the instinct of the greatest scientific intuition of the twentieth century and try to find a way to a new science in which Einstein's "painful resignation" will not be necessary?[68]

This is indeed one of the central themes that I have expounded in this book. My suggestion is that we can only understand time in terms of both the existence of the present (the "now") and the distinction between past and future once we achieve a unified understanding of existence that includes a primary role for the mind and sentient awareness. This will need to include the realization that mind cannot arise as an emergent property of matter or time but, instead, that it is fundamental and has no beginning or end.

Smolin gives a clear description of the problem, which is often evaded when many scientists consider the question of consciousness:

> By the problem of consciousness I mean that if I describe you in all the languages the physical and biological sciences make available

to us, I leave something out. Your brain is a vast and highly interconnected network of roughly 100 billion cells, each of which is itself a complex system running controlled chains of chemical reactions. I could describe this in as much detail as I wanted, and I would never come close to explaining the fact that you have an inner experience, a stream of consciousness. If I didn't know, from my own case, that I'm conscious, my knowledge of your neural processes would give me no reason to suspect that you are.

What's most mysterious, of course, is not the content of our consciousness but the *fact* that we're conscious. Leibniz imagined shrinking himself down and walking around inside someone's brain the way he might walk around inside a mill (these days we would say "a factory"). In the case of the mill, you could give a complete description of it by describing what someone walking around inside it would see. In the case of a brain, you couldn't.[68]

I would add that if you include the electrochemical activity of the brain, you could form a complete description of it as an object, but that description, though it might make reference to some of the contents of our perception, would not explain them, and more to the point, it would come nowhere near to explaining the fact that we exist as sentient entities in order to perceive at all!

Is it merely coincidence that the two elements of our experience that are most difficult to define in a noncircular (self-referential) manner are the flow of time and consciousness?

We can't define consciousness in a reductionist manner any more than we can describe the "qualia" of our experience. As an example, how would you describe what the color red looks like without simply listing things that are red in color? However, someone who has seen the color red knows how it appears to him or her, and someone who is conscious experiences what it is like to be conscious and also seems to experience a subjective flow of time.

The notion of the flow of time implies an inherent difference between a fixed past and an indeterminate future. Philosophers such as David Chalmers have pointed out that notions of consciousness as an emergent phenomenon have (among other logical flaws) the difficulty that they don't explain why consciousness is necessary if it is merely a passive bystander in behaviors that (in a reductionist model) would be determined by physical processes in the same way whether or not they were consciously observed. However, along with our experience of being conscious, we also have the sense that we are somehow in control of our actions. Indeed, our language and moral and legal systems are all based on the presupposition of personal responsibility for our actions in the sense that we could have done otherwise and that we act with free will. To quote Schrödinger again:

> My body functions as a pure mechanism according to the laws of nature. Yet I know, by incontrovertible direct experience, that I am directing its motions, of which I foresee the effects that may be fateful and all-important, in which case I feel and take full responsibility for them. The only possible inference from these two facts is, I think, that I—I in the widest meaning of the word, that is to say, every conscious mind that has ever said or felt "I"—am the person, if any, who controls the "motion of the atoms" according to the Laws of Nature.[2]

4. The Arrow of Time

The term *arrow of time,* as we have seen, was coined by Arthur Eddington (see page 75) and relates to the observation that disorder, as measured by the amount of entropy, always increases in a closed system.[28]

A strong case can be made that there could be a statistical explanation for this in that the probability of a more ordered state arising spontaneously is vanishingly small. However, this argument should work the same in both "directions" in time.[76] The enigma in need of explanation is why the universe began in such an amazingly extreme state of order such that there could be an arrow in the first instance. It is my contention that if the universe had begun from nothingness, then this would explain the low-entropy boundary condition. Actually, the entropy state of true nothingness is zero!

My suggestion is that the existence of free will is implied by all four aspects of time.

The "now" is the product of sentient awareness. The awareness of options is possible because there exists freedom of choice between these options.

The "origin" of time would then be the actualization of the choice to exist in separation by sentient beings. Space, time, and matter are implicit consequences of the choice to make separation happen.

The "flow" of time is the dynamic interface between what is and what can be (i.e., between actuality and potentiality). Free will is that which can convert potentiality into actuality.

The "arrow" of time indicates our origin from a state of union and indicates that perhaps the initial impetus that spawned the big bang was our choice to experience separation, which then led and continues to lead to an inexorable increase in disorder and entropy, as described by the second law of thermodynamics.

However, if it is free will that is responsible for this seemingly inescapable degeneration into chaos, then one might consider that there might be another way: a state of existence where there is no decay and where there is perfect knowledge, compassion, and peace; a state perhaps glimpsed as the "light" by some near-death experiencers.

Perhaps this is the same light that formed the image on the Turin Shroud, and perhaps the man whom the Shroud once wrapped told us about and demonstrated a modus vivendi for achieving that state and *becoming* that light without the need for dogma or religion but simply through uniting rather than dividing and following the most reasonable suggestion of all—to love our neighbor as ourself, because within that light, our neighbor *is* our self.

Five Points
What I Am Not *Saying!*

My reason for writing this chapter is to try to avoid misunderstandings. I have often seen comments from people about books in which they have said that the author is saying "x" or the author is from the "y" school of thought. Following these, authors will write an article or another book explaining that that is not what they said. I thought I would preempt this by guessing at ways that what I have said might be misconstrued and attempt to clarify. With this in mind, below are five examples of what I am *not* saying.

1. I am neither advocating the position that existence is some type of dream nor that it has no objective reality.

In suggesting that consciousness is fundamental to all reality, including physical reality, I am suggesting that consciousness is the cause of physical reality and not that reality is imaginary. As human beings, we clearly have physical bodies that are subject to decay and disease and to pain and suffering. These cannot be conveniently explained away as merely the product of imagination.

My suggestion is that the physical universe in all its myriad expressions is the very real product of choices made through the expression of free will. Some might ask how that is possible if conscious beings with free will only emerged more than thirteen billion years after the universe began. Well, I would question the assumption behind this question.

I have argued that the emergence of "somethingness" from the no'thing'ness from which the big bang occurred would be a product of a self-willed choice that propelled sentient being into separation, thereby creating space, time, and matter. Once separation of sentience into individuals began, then there would have been some unevenness or inhomogeneity in the primal universe that then became the focus for galaxy formation and so on. If choices made in free will equate to primary causation and all else is secondary causation, then it could follow that the vast majority of the explanation for why the universe has taken the form that it has would be due to secondary causation.

If I might repeat an example from chaos theory, then it might be easier to see why this is so. You may be familiar with the notion that weather systems are so finely balanced that even a butterfly flapping its wings could disturb the system sufficiently as to cause a storm thousands of miles away. If we imagine that the butterfly might have chosen to flap its wings, it is not hard to imagine that it would not have chosen to create the storm . . . but it did cause the storm. Of course, the butterfly was not the only cause; there would have been a multitude of other factors that caused the weather system to be poised with that particular set of parameters such that the flapping wings were sufficient to tip the balance. My point, though, is that the physical universe, once it appears, is a chaotic system and so depends on an intricate chain of cause and effect. If free will is, as I have argued, synonymous with primary causation, then points of choice must ultimately be the (usually indirect) cause of all that happens.

2. I am not advocating creationism.

In suggesting that the universe appeared out of no'thing'ness as a result of a decision to exist in separation, I am not postulating a "creator god" who "lit the fuse" to start the whole process off!

I have suggested that the existence of a physical universe containing consciousness implies a continuum between matter and mind, with matter, space, and time arising as a result of the restrictions we have taken on through limiting our capacity to know. In viewing the world through self, we limit our capacity to know, as I can only see through my eyes and you through yours. These limitations imply difference, which implies separation, which itself relates to the existence of space, time, and matter. My view is that this is the *explanation for why and how space, time, and matter exist at all!* I have argued that if, as Schrödinger suggested, time is a product of mind, then mind must be eternal. If the universe began from no'thing'ness, from a state of union from which separation emerged, then, as sentient beings, we must have our roots in that original state of union. Since time began with separation, it stands to reason that the source state is beyond time and therefore unchanging, containing all potentialities within it. Also, it would logically consist of sentience beyond limit, and so it would be omniscient. Actually, it fulfills many of the criteria that people might imagine would be used to define what they call God. However, it would not fit the description of the anthropomorphic "old man in the sky," meting out rewards and retributions based on some kind of insecure need to be appreciated and worshipped!

The notion of a directive creator god is therefore, perhaps, as many scientists have argued, no different in kind from a myth that says that the world is supported on the back of a turtle. This raises the question of what is supporting the turtle, to which an answer might be that it is on the back of another turtle and that it is "turtles all the way down." After all, one could reasonably ask who or what created the creator and so on.

I would suggest that potentiality is an implicit property of no'thing'ness, and if nothingness is the ground of all being, then we don't need to appeal to notions of creator beings to account for primary causation. We don't need a chain of turtles! Every time we make a choice, that choice is itself primary causation. As I postulate nothingness or, rather, no'thing'ness, as the ground of all existence and I identify free will (i.e., primary causation) as a natural property of the state of no'thing'ness, I hope you will agree that this is a "turtle-free" approach to trying to understand existence!

I have suggested that if we were to use the term *God* or *godhead* to describe the ultimate pole of all union, then it would make more sense to see it as the center of all potentiality rather than as the actualizer of that potentiality. Any *action* implies time and change, as there is a *reaction* that defines the one who acts as being different *after* the action from how he or she was *before*. Any temporal act therefore defines the one who acts as being temporal and specific and therefore distinct from the pole of all union in no'thing'ness. My suggestion is that the actuation of the potential for separation arises through the choice of sentient beings.

How can they make the choice to exist if they have to exist in order to choose?

The answer is that they (or we) existed already, but our state of separation existed as *a potential*. In actualizing that potentiality, we merely created a new reference frame of subjectivity in separation. We already exist(ed) in the whole as the whole, and in choosing to separate, we did not and could not diminish the whole that exists beyond time and *still* contains us in our unlimited aspect.

So . . . what am I proposing to be the nature and properties of this preexistent state of union beyond space and time? Well, it is not union in the sense of bringing together *things* in the manner of a jigsaw puzzle. It is a union of all sentience. Perhaps the best way to relate to this from within our existence as human beings is through what is generally termed empathy and *love*. This is not meant in some nebulous or romantic sense but as a logical surmise regarding how the union of sentience can be viewed from within separation. "Love thy neighbor as thyself" rings true if, as I have argued, our neighbor ultimately *is* our self as we all have our roots in the state of perfect union. To be beyond time, this love from the godhead would by definition be completely unconditional because conditions imply contingent changes and it has no need of change. After all, if it changed it would have to be temporal!

3. I am not advocating any particular religion.

If we accept the evidence (see pages 9, 15–19, and 25) that the man of the Shroud is Jesus of Nazareth, then it is interesting to bear in mind that when questioned about his role in life (John 18:37), he was reported to have replied that he had been born to "bear witness to the truth."

It's interesting that this does not carry the implication of the founding of any religious institutions. In fact, if you look at his interactions with the religious institutions of his day, a case could be made that he seemed to be very derisory about the tendency of such institutions to obfuscate sense and reason and to indulge in hypocrisy and materially motivated agendas. I would suggest that if we are to make a fresh start at trying to understand his teachings and his life, then we might do well to consider that we don't need to be tied to the interpretations of the institutions that *claim* to represent him. Perhaps it might also be worth looking at some of the gnostic texts that were excluded by the early church.

This book is about the implications of the image on the Shroud, and therefore a lot of it centers on the teachings and life of Jesus of Nazareth as all the evidence points to him as being the man whose image we see on the Shroud. This does not in any way belittle the magnificent contributions to humanity made by other inspiring teachers such as Siddhartha Gautama (the Buddha), Zoroaster, Guru Nanak, the Judaean prophets, Muhammad, and many others.

4. I am not claiming that the notion of miracle needs to be invoked to explain the Shroud image.

I have throughout this book taken many opportunities to expound on the idea that physics as it stands is incomplete. It has demonstrated through quantum mechanics that there is an essential and inescapable role of consciousness in physical reality, but it has explained neither the origin and source of consciousness nor *how* consciousness and matter are related. I have postulated a mind-matter continuum in which matter, space, and time are the expressions of consciousness, which has taken on limit and therefore separation. I have argued that in taking Bohm's point about matter being merely frozen light seriously, we need to consider the nature of light if we are to understand matter. We can see through Einstein's equations that the speed of light implies an underlying unity to the constituent points in space and time and that those points are defined by matter. In other words, what we now perceive as separation, location, and the apparent flow of time are all results of the interaction of consciousness with light that has been frozen as matter. If, as I have suggested, the physical universe emerged as a result of a choice to separate, then that choice would have frozen light, and our apparent existence "in" material bodies would therefore be the result of that choice.

If, as our day-to-day experience and also quantum theory (see page 126) suggest, we do actually possess freedom of will, then that would imply that perhaps the degree of freezing of light is under our control through how our thought is directed, as expressed by how we live. If it is the choice to separate that freezes light, then the genuine and consistent choice to unite through living completely according to the principle of *love thy neighbor as thyself* would be able to unfreeze the light, such that a physical body could "shine." The second law of thermodynamics is a description of the fact that every "thing" breaks down and decays back toward nothingness. Our fundamental status as sentient beings is not to be "things." Empathy is simply the recognition that you are not a thing and that other people are also not things but have limitless value and significance. Perhaps a deep recognition of this fact might give us the strength to forgive and to "love our enemies." It is my position that the evidence of the Shroud image suggests that human beings do have the potential to achieve this completely. During near-death experiences, people often experience everything that they have done to others as being done to them. It seems to me that what the man of the Shroud was advocating was to extend that deep empathy into our daily lives such that we might truly love our neighbor as ourself. In doing so, perhaps we might embody the "light" that people see in NDEs while we are alive so that we do not need to wait until we are near death to appreciate its meaning.

At this point, I would remind you of the evidence (see chapters 1 and 2) that the Shroud image may have been formed as a result of a short, intense directional burst of radiant energy from the body of the man of the Shroud. This was possible because of the way he had lived his life. I hope it is clear from this that my position is that the formation of the Shroud image was the *completely natural* result of how the man of the Shroud had lived and did not require any "divine intervention" in the form of a "miracle."

5. I am not advocating dualism.

I am suggesting that mind and matter are not a duality but a continuum, like light and shadow. Shadow cannot exist without light, but light can exist without shadow. In the same way, matter is seen in my thesis as an emergent, real product of mind that exists as a result of a choice of sentient being to exist in a state of separation, such as the separation that exists between all of us human beings as sentient individuals.

EPILOGUE
Natural vs. Artificial

Your mind is not something that could be merely a product of the gelatinous substance that is called your brain. An analogy that I use to illustrate this point of view is that of a radio. When you switch your radio on and hear a song, it is not usual to believe that the radio is the composer, interpreter, or singer of the song. If I am right, then the currently fashionable aspiration of someday striving for immortality by "uploading" our consciousness into machines can be seen as a deadly and ill-informed denial of our true nature as human beings. As I argued at the start of this book, if we fail to see what distinguishes us as sentient beings from mere material objects, then we might lose the safeguards that save us from destruction. The continued existence of these safeguards might come to depend on the tacit recognition of the limitless value of each and every human being as rare oases of living awareness and free will in this perhaps otherwise empty desert of a material universe. It is the widespread recognition of this grandeur and immensity of human potential, however, that could be the essential first step to achieve peace on Earth. Without it, our species might be doomed to annihilate itself, as perhaps numerous other intelligent (and unwise) civilizations throughout the universe have before us.

Recent astronomical discoveries suggest that there might be billions of Earth-like planets in our galaxy alone. Fossil evidence suggests that life on Earth began very quickly once conditions made it habitable for living organisms. That being so, we would expect that many other planets would also be inhabited, and if even a small proportion of these included intelligent and friendly life-forms, then why do we not see evidence of these? Could it be that they, like us, followed the pathway of technological development and then destroyed themselves through environmental change or war? Perhaps they might even have simply lost sight of what it is that distinguishes sentient living being from "artificial intelligence." If that were the case, then they might have even sought to perpetuate their existence through artificial intelligence and "mind uploading." These "developments" would, if I am right about the nature of mind, be doomed to fail and would certainly be counterproductive. In December 2012, the University of Cambridge in England announced the formation of the Centre for the Study of Existential Risk, a project to investigate the risks that developments in technology might pose to the continued existence of our species.[77]

The following is a quote taken from the University of Cambridge web page of Lord Martin Rees, a member of the center and the current Astronomer Royal.

> Many scientists are concerned that developments in human technology may soon pose new, extinction-level risks to our species as a whole. Such dangers have been suggested from progress in AI, from developments in biotechnology and artificial life, from nanotechnology, and from possible extreme effects of anthropogenic climate change. The seriousness of these risks is difficult to assess, but that in itself seems a cause for concern, given how much is at stake.[78]

Lord Rees also has implied that it would be expected that if we are not careful, humanity may be replaced by machines and that if we encounter signs of extraterrestrial life, they would probably be dead mechanical echoes from a once-living species that has been replaced with machines. He was interviewed by *BBC Science Focus* magazine in April 2016 and asked, "What might an alien signal look like?" He replied in the following way:

> We would be looking for a signal that was not natural. It might have a very narrow bandwidth, or be pulsed. But even if it were clearly artificial, that wouldn't mean it was a message that we could decode. My guess is that it wouldn't come from a civilisation of organic beings but from some hyperintelligent machine. Let's think what has happened on Earth, and what may happen in the future. Technological civilisation emerged after 4.5 billion years, but within a few centuries it may have been surpassed by machines that spread into space—and they will have billions of years to develop further. So it may well be a fairly brief period in Earth's history during which intelligence is dominated by organic creatures, and a much longer future when it's dominated by machines. This therefore suggests that, if alien intelligence has emerged on another planet via a similar route to what happened on Earth, we are unlikely to catch it in the brief organic stage—we're far more likely to catch it in the far longer post-organic stage.[27]

So we have come full circle, back to a subject mentioned in the introduction to this book. Could our headlong rush into technology and materialism have made us lose sight of our humanity to the extent that the very continued existence of the human species is now under threat? Look out at the night sky and you might see billions of stars so far away that the distance defies imagination, and yet all of them are contents of your perception. Your mind is greater in scope than the entire physical universe. More to the point, look at someone you really care for, perhaps your child. Is that person really just a pile of atoms arranged in cells, or is there something looking out from behind his or her eyes as well, something that is greater than all materiality?

The aim of this book has been to invite you with me on a journey to investigate what that "something" might be. Many people who have had near-death experiences recall a light from which they experience a sense of perfect knowledge and unconditional love. This book has been an attempt to understand the nature of humanity in reference to this light and to explore the possibility that the evidence from the image on the Turin Shroud might show that this light is not an "artifact of a dying brain" but is the *fundamental basis of existence itself.* If so, then at least an ember of that light is in you too, as it is in all sentient beings, and just as that light is in you, your full potential is contained in that light. In a very real sense, then, you are the light of the Shroud, and the image is there because of you and for you.

When I give lectures about the Shroud of Turin, I am frequently asked whether I believe the man of the Shroud was who he said he was. I like to answer the question with a question. Do you believe you are who he said you are?

APPENDIX

Scientific Investigation of the Shroud

CARBON DATING

Carbon dating is one of several techniques used to estimate the age of historical artifacts. It works on the basis that there are three naturally occurring varieties, or *isotopes,* of carbon. The isotopes are distinguished by the number of particles contained in the nucleus at the center of the atom. All carbon atoms have six protons, and around 99 percent of them also have six neutrons. These atoms are known as carbon 12, having twelve particles composing their nucleus. Around 1 percent of naturally occurring carbon is carbon 13. These atoms have six protons and seven neutrons.

Both carbon 12 and 13 are known as *stable isotopes.* They are described as stable because they do not undergo radioactive decay. However, there are also trace amounts of carbon 14. Carbon 14 is present in tiny amounts, around one part in a million-million. It is known as a *radioisotope* as it is (weakly) radioactive. This is because it is unstable and over a very long period of time will undergo radioactive decay. Carbon 14 has a half-life of around 5,730 years, which means each atom has a one-in-two chance of undergoing radioactive decay within this timescale. Carbon 14 mainly arises from a nuclear reaction that occurs in the upper atmosphere. *Cosmic rays,* which consist mainly of high-energy protons, collide at high speed with molecules in the upper atmosphere, which results in the secondary production of neutrons. These neutrons collide with atoms of nitrogen, which, in turn, causes a reaction that converts the nitrogen to carbon 14. Carbon 14 is, as we have seen, unstable and decays to convert gradually back to nitrogen 14. However, an equilibrium occurs due to the replenishment of carbon 14 in the atmosphere (mainly in the form of radioactive carbon dioxide). Plants incorporate atmospheric carbon (from carbon dioxide) through *photosynthesis,* and so while the flax plants from which the linen of the Shroud was made were alive, the ratio of carbon 14 to carbon 12 would have remained constant. However, once an organism dies, metabolism ceases, and so the proportion of carbon 14 gradually declines in a predictable manner. It is this gradual decline in carbon 14 by the process of radioactive decay that allows us to use the *isotope ratio* to calculate the approximate age of an artifact.

In 1989, the journal *Nature* published a paper reporting on a carbon-dating study that had been performed on material taken from the Turin Shroud.[79] The result of this study was widely reported by the media at the time as suggesting that the Shroud originated between 1260 and 1390.

I was a student when the report was published, and as I was studying for a science degree that I did as an extra degree during my medical studies, I used to read *Nature* each week. I remember remarking on a certain discrepancy in the carbon-dating report. There was a table of data that compared the findings of the three carbon-dating labs and set this against their quoted error ranges. According to this, when comparing the data from the labs, the hypothesis that they had dated samples with the same age could be rejected with 95 percent confidence! Twenty-two years later, I spoke at a scientific conference on the Shroud at ENEA, the Italian National Agency for New Technologies, Energy, and Sustainable Economic Development. One of the other speakers there was a Marco Riani, who presented a fascinating paper in collaboration with a statistician from the London School of Economics about the statistical analysis of the 1988 carbon-dating study.[80] This confirmed my suspicion that the statistics showed a flaw in the sampling methods. The sample taken was not homogenous in age and therefore was unlikely to be representative of the rest of the Shroud.

The radiocarbon labs of Oxford, Zurich, and Arizona had each been given samples taken from a single seven–by–one centimeter strip of cloth. This strip had been cut from part of the cloth that would have been expected to have been among the most contaminated through history as it was taken from a corner. The cloth used to be handled by the corners, and this was also where it sustained the most damage apart from the parts that were burned in a fire in 1532.

Around the start of the twenty-first century, a husband-and-wife team in the United States, Sue Benford and Joe Marino, were studying some photographs of the Shroud when they noticed that it looked like there were two distinct appearances of the cloth around the area where the carbon-dating sample was taken. They did some investigation into this and found that there was a technique of cloth repair that had been used in France and also by Margaret of Austria's master weavers. Margaret of Austria (of the House of Savoy) was a custodian of the Shroud in the sixteenth century. This was known as an "invisible reweave" and is a technique that is still used to this day. It involved repairing a damaged cloth using extraneous material that was then spliced into the original material in such a way as to preserve the original weave pattern. Benford and Marino speculated that perhaps this had been done in the case of the Shroud, which thereby perhaps accounted for the carbon-dating result—a result that was anomalous compared with all the other data, and which suggested a much earlier date for the cloth.

One of the scientists who had been a member of the 1978 STuRP team (see chapter 1) was Raymond Rogers, a chemist doing research at Los Alamos National Laboratories. He responded to Benford and Marino's work by expressing his annoyance that these "fringe" researchers were suggesting that the carbon dating had been wrong. Prior to the carbon-dating, Rogers had believed that the Shroud may well have been of the first-century in origin, but after the carbon dating, his opinion had changed. He still could not account for how a putative "medieval forger" could have produced something with the forensic appearances of the bloodstains or the characteristics of the Shroud image, but he felt that the carbon-dating result should have been accepted. He still had some samples of the Turin Shroud in his possession, including parts of the Raes fragment, which had been cut by Professor Gilbert Raes in 1973 from an area adjacent to what would later become the carbon-dating sample. He said that he had the material to "prove [Benford and Marino] wrong in five minutes" by demonstrating that the carbon-dating area was indeed representative of the whole cloth and had an equivalent chemical composition. He then looked at the microscopic and chemical characteristics of the two samples (the Raes sample and another sample from a different part of the Shroud) and came to an astonishing conclusion: Benford and Marino were right! He found that samples taken in the vicinity of the carbon-dating sample were different from samples taken from elsewhere on the cloth and seemed to have cotton fibers interposed on the linen. They appeared to have been dyed to give them the same color as the cloth, and this was also anomalous compared with the rest of the cloth, as the STuRP scientists had found that the main body of the Shroud, including the image-bearing parts, were not colored by any dye, paint, or pigment. One observation that he found most convincing was that one of the fibers appeared to show splicing together of two completely different types of material.[81]

An interesting feature of Benford and Marino's reweave hypothesis[82] is that it would account for the discrepancies between the apparent carbon-dating age of the adjacent portions of the strip that was sent to the three laboratories,[80] as they could have had different proportions of original Shroud material and extraneous "repair" material. The original plan for the carbon-dating study had been to use seven laboratories and two different carbon-dating methods and to take samples from more than one part of the cloth in an attempt to ensure that the parts that were carbon-dated were representative of the rest of the cloth. It had also been suggested that chemical studies be performed on the samples to ensure that they were representative of other parts of the cloth. In the end, all of these original plans were thrown out. As it happened, the representatives of the radiocarbon labs refused to allow any of the STuRP scientists to be involved in the decision about where the samples should be taken from. The STuRP scientists were not amateurs but were a collection of neutral experts brought together from many fields, including, for example, a chemistry expert from Los Alamos and people who had been involved in the NASA space program. The STuRP scientists had, as I have previously mentioned, conducted a very extensive study on the Shroud, but their advice was deemed irrelevant.[83,84] To many scientists, all of this implies that the findings of the original carbon-dating report cannot be considered in isolation from all the other evidence, which suggests a much earlier origin for the cloth. The carbon dating does not carry sufficient scientific weight as it has not been established that the sample tested was representative of the cloth from which it was taken. Professor Christopher Ramsey was part of the original carbon-dating team, having been working at the Oxford Radiocarbon Laboratory as a postgraduate student in 1988. He now runs the lab. He is still not convinced of the Shroud's authenticity, and for now, as far as I know, is still sticking to the medieval date for the Shroud, but despite this he said the following when interviewed as part of David Rolfe's BBC Television documentary on the Shroud:

With the radiocarbon measurements and with all of the other evidence which we have about the Shroud, there does seem to be a conflict in the interpretation of the different evidence. And for that reason I think that everyone who has worked in this area, radiocarbon scientists and all of the other experts, need to have a critical look at the evidence that they've come up with in order for us to try to work out some kind of a coherent story that fits and tells us the truth of the history of this intriguing cloth.[85]

THE BLOODSTAINS AND THE BODY IMAGE

In 1978, the STuRP team applied sticking tapes to the blood marks and the image marks. They sent the tapes to Professor Alan Adler of Western Connecticut State University. Adler was a chemistry professor and an expert on chemicals in the blood called porphyrins. By the detection of haem derivatives, which are bile pigments and proteins in the globules that had been removed from the bloodstain areas, Adler established the presence of whole blood on the cloth.[5,6] Adler analyzed the fibrils removed from the image areas. He found that there were no paints, pigments, stains, or dyes causing the coloration of the image fibrils, but instead there was a chemical change in the fibrils that consisted in dehydration and oxidation to cause the yellowing of the fibrils. To quote Adler and John H. Heller from their 1981 *Canadian Society of Forensic Science Journal* article referenced above:

> There is no chemical evidence for the application of any pigments, stains or dyes on the cloth to produce the image found thereon. The chemical differences between image and non-image areas of the cloth indicate that the image was produced by some dehydrative oxidative process of the cellulose structure of the linen to yield a conjugated carbonyl group as the chromophore. However, a detailed mechanism for the production of this image, accounting for all its properties, remains undetermined.[5]

This chemical change that was identified in the image-bearing fibrils is a form of accelerated "aging" of the cellulose in the fibrils that has very similar properties to that which causes the yellowing of paper that has been exposed to sunlight. Among other methods of study, the bloodstains and image were examined by direct inspection, using photomicrographs, and by analysis of the material removed on the tapes. Several contrasting properties were found between the two types of mark on the cloth.

1. The bloodstains soaked right through the cloth, while the image discoloration is purely a surface phenomenon.

When the backing cloth was removed, it was seen that the bloodstains were visible on the reverse side of the cloth but the image marks were not. Also, if the shroud is viewed in light transmitted through the cloth, one can see that the bloodstains are visible but the image marks are not. More specifically, the STuRP researchers found that the discoloration was only on the outermost superficial fibrils, which are each only around one-hundredth of a millimeter thick (i.e., much less than the thickness of a human hair). Despite the tiny diameter of the fibrils, the discoloration that constitutes the image is tinier still, as only the surface of the fibrils is affected. So, the discoloration in the superficial part of the fibrils is thinner still—around 200 nanometers in thickness (i.e., around one five-thousandth of a millimeter only). Also, where a discolored fibril passes underneath another fibril, the discoloration does not continue in the part that is obscured by the overlying fibril. Where a fibril follows the natural curve of the thread, only the outermost aspect of the fibril is discolored. This shows that there is no wicking as would have occurred if the color change had been caused by a liquid applied to the surface of the Shroud.

2. The bloodstains can be clearly seen to have been caused by an extraneous substance with which the cloth had made contact, but the image marks consist of a chemical change in the fibrils, only with no substance having been added to the cloth to form the image.

Photomicrographs (photos taken by a microscope) of the bloodstain areas clearly show that the fibers have been cemented together, and debris can be seen between the fibrils (see plate 6), but equivalent magnification photos of the image areas clearly demonstrate pristine fibrils, some of which appear yellow rather than white, and they are not cemented together. No debris is visible on the image photomicrographs (see plate 5).

3. There is no apparent variation in shading of the yellowed fibrils.

The amount of blood on the stains varies in density between different parts of the stains, but the discoloration process in the body image is an all-or-nothing phenomenon. The fibrils are either yellow or white. The variations in intensity of the image consist not in shades of color but in the number of fibrils that have been changed. This is similar to the way black-and-white newspaper photographs are shaded. They are made of uniformly colored dots, and the darker parts simply contain more dots per unit area.

4. There is no yellowing of the fibrils underneath the bloodstains.

There are two very important conclusions that follow from this. The first is that the image-formation process happened *after* the bloodstains were formed, and the second is that the presence of the blood in some way prevented the change in the fibrils that caused the image to form. In some way, the blood "shielded" the underlying fibrils from being affected by the image-formation process. I have argued that this has great significance when possible explanations for the image-formation process are explored. Again, research by the Shroud of Turin Research Project group supports the observation that there is no image under the bloodstains.[8]

5. The formation of the bloodstains and the image appear to have occurred through very different processes.

The bloodstains appear to have been formed by a contact mechanism between the cloth and a dead body of a man within the first two hours after death.[14] The man appears to have been tortured and then crucified in a vertical position prior to being laid out supine on the cloth to account for the dorsal bloodstains from the back of the body. Approximately one-half of the cloth also appears to have been folded over the vertex of the head so as to drape over the front of the body, and this caused the bloodstains to appear on the cloth from the front aspect of the body.

Dr. Gilbert Lavoie has demonstrated that although the bloodstains were formed by a contact mechanism from a cloth that was draped over the body, when we see the cloth flattened out, the stains are seen to be moved outward.[14] His simple experiments demonstrate how blood from the face, for example, becomes superimposed on the image of the hair. There are also some bloodstains that appear to be away from the body image altogether. One such is the "off-image" elbow stain. Lavoie demonstrated that this is consistent with a flow of blood along the arm from the wrist while the man of the Shroud was suspended in a position of crucifixion. The draping and tucking of the cloth around the body upon its being taken down from the cross could then account for the position of the stain. This is further strong evidence that the cloth once contained the recently deceased three-dimensional body of a victim of Roman torture and crucifixion.

At this point, it might be helpful to take stock of certain implications that arise from the observed differences between the bloodstains and the body image. The off-image elbow stain and other off-image stains, such as those outside the edge of the hair image, provide evidence that the configuration of the body and the cloth could have changed at some point between the formation of the bloodstains and the appearance of the body image!

We have seen that, forensically, it can be shown that the bloodstains were formed by a contact process soon after death when the body was placed on the Shroud, the Shroud then being folded over the head (see plate 11). What then were the positions of the body and the cloth when the image formed?

When a body is laid out on its back, one sees flattening of the posterior aspect of the body, particularly the buttocks and the calves. This is consistent with what we see on the Shroud, as there are wide areas of contact between these parts and the Shroud, as seen from the fact the bloodstains have transferred to the cloth over a wide area.

However, when we study the body image, we see that flattening is not evident in these areas on the body image. Some people have argued that this is because of rigor mortis, which they say caused these areas to be rounded due to intense contraction of the underlying musculature. However, observations of bodies in rigor mortis confirm that rigor does not abolish obvious signs of flattening. This is because rigor mortis only affects muscle and not the skin, subcutaneous tissues, or the layer of fat that exists between the skin and the muscle, even in slim people. Interestingly, rigor mortis is a temporary phenomenon that normally resolves within twenty-four to thirty-six hours.

As we have seen, the image contains three-dimensional information encoded in the intensity of the image, which means that those parts of the body such as the face and the hands that would have been closer to the cloth are associated with a more dense pattern of alteration in the fibrils than those parts of the body that would have been farther from the cloth. Also, the distortion associated with the draping of the cloth that we see in the pattern of the bloodstains is not visible in the image. This is why the bloodstains from the face are superimposed on and around the image of the hair. There is also image present of those parts of the body in recesses that would not have been in contact with a draped cloth. This supports those theories of image formation that do not involve contact between the body and the cloth, such as the hypothesis that the image was formed by a burst of radiant energy from the dead body that the Shroud wrapped.

As we do not see flattening of the muscles at the back of the body, we can infer from this that at the time of image formation, these muscles were not compressed by the weight of a body laid on its back. However, as mentioned earlier, the pattern of the bloodstains is consistent with the transfer of the blood to the cloth occurring when a dead body was laid horizontally on the cloth. This reaffirms Lavoie's observation that it is as though there were two different events recorded on the cloth. The first was the transfer of the bloodstains to the cloth, which occurred when a dead body was laid out on its back on the cloth and the cloth was turned over the top of the head and laid out over the front of the body. The second was the image formation, which clearly occurred after the bloodstains had already formed, as there is no image underneath the bloodstains. What position was the body in at the time the image formation occurred?

We have seen that at the time of image formation, the back of the body was not flattened under the weight of a body as one would usually expect if the body had been lying on its back. Could there be any other clues to the position of the body at this time?

If one looks at the position of the hair, then something very interesting becomes apparent. Lavoie pointed out that if a body (particularly one with long hair) is laid out face upward, then the hair will fall backward and spread out behind it.[14] However, in the case of the man on the Shroud, one can clearly see that the hair is hanging down on his shoulders.

This suggests that at the moment of image formation, the body was not horizontal but vertical!

Next, one would naturally look at the image of the feet, as under normal circumstances human bodies are standing when they are vertical. However, in the Shroud image, the feet are pointing downward and the left foot is at a different level from the right one, which suggests something even stranger.

It would appear that at the time at which the image formed, the body was upright and *suspended in the air* as though it had become virtually weightless and had somehow moved into a vertical position in which it was "floating"!

This may well sound like a bizarre conclusion to form from a piece of cloth. Actually, it is not a conclusion but a working hypothesis. At this point, perhaps, I should remind you that forensic experts who have studied the Shroud have reached a consensus from the evidence of the blood and serum stains on the cloth that it was used soon after his death to wrap the body of a man who had been whipped, tortured with a cap made from sharp objects such as thorns, and then crucified. To date, nobody has been able to account for how the image of this corpse appeared on the cloth in the form of surface changes to linen fibrils with a thickness of only one five-thousandth of a millimeter, conveying accurate anatomical and pathological information, and with photographic negative and distance-coded features. All the evidence is consistent with the hypothesis that the man of the Shroud may well have been the historical Jesus of Nazareth. It is interesting to note that the evidence that the body was upright and suspended in the air only became apparent with nineteenth- and twentieth-century technology, and the evidence that the features of the image are consistent with a short, intense burst of radiant energy from the body that the Shroud wrapped only became apparent through twentieth- and twenty-first-century technology. These features of the bloodstains and the Shroud image could not have been apparent to people in medieval times.

IMAGE-FORMATION HYPOTHESES

There are many suggestions that have been made over the years as to the mechanism of formation of the body image on the Turin Shroud, and many of these can be dispensed with very simply by readily demonstrable evidence.

1. The Painting Hypothesis

A painted image would be readily apparent upon direct inspection, and various photographic and chemical analytic techniques have been applied to the Shroud, particularly in 1978 by the STuRP team (see chapters 1 and 2). Tiny traces of paint have been found on the cloth, but at an insufficient level to form an image, and also they are not concentrated in the image-bearing areas of the cloth.[5,8]

It is thought that when artists in the past painted copies of the Shroud, it would have been their custom to press their paintings against the Shroud. They would perhaps have believed that this would transfer some form of "blessing" to their images, but the contact would also have transferred minute amounts of paint to the Shroud. If one inspects photomicrographs of the image-bearing regions of the Shroud (plate 5), one can immediately see marked differences from what would be seen if the linen had been painted. The paint medium would have soaked through the fabric of the cloth, and also would have matted the fibers together and left particulate matter visible on them. As we have seen, the image does not consist of anything that has been added to the cloth, but instead Adler's extensive chemical testing showed that the discoloration of the surface fibrils of the linen in the image-bearing areas results from oxidation and dehydration.[5] Professor Luigi Garlaschelli of the University of Pavia claimed to have reproduced the properties of the Shroud image using applied pigments, but when his handiwork was inspected microscopically, it showed a completely different distribution of coloration from that of the Shroud image, as the Shroud image fibrils show uniform discoloration along the fibrils and similar discoloration when the image-bearing fibrils are compared to each other. The intensity of the Shroud image is dependent on the number of fibrils discolored per unit area, while Garlaschelli's image shows patchy coloration on the fibrils and variation in intensity along the fibrils, depending on how much pigment was applied.[86]

Also, one of the unique features of the Shroud image is that because the formation of the image was confined to the uppermost fibrils, the image is not visible when the Shroud is lit from behind. Conversely, Garlaschelli's attempt at a replica produced an image that was clearly visible when illuminated in this way.[87]

Joe Nickell, a magician and professional skeptic who also holds a Ph.D., attempted to reproduce a Shroud-like image using iron oxide, and, when his work was examined microscopically, one can again clearly see a type and pattern of discoloration that is completely different from the one on the Shroud. It bears repeating that the STuRP team, in their multidisciplinary examination of the Shroud, found only miniscule amounts of iron on the Shroud body-image areas (although, as one might expect there was iron present in the bloodstains). The amount of iron present was not higher in the image-bearing areas compared with the non-image-bearing areas and was not high enough to produce a visible discoloration.[5]

2. Contact Processes

One of the obvious features of the Shroud image is that the recesses of the body surface are clearly visible. Processes that rely on contact to form an image will leave imprints of the prominent areas of the body and will mask those areas that are recessed and therefore not in direct contact with the cloth. This would be the case whether the image were formed by contact with an actual corpse or with a statue. Also, draping the cloth over a body or either a bas-relief sculpture or statue would cause an apparent gross distortion of the image when the cloth is viewed in a flat planar aspect.

Vittorio Pesce Delfino of the department of zoology at Bari University attempted to form a replica of the Shroud image using a combination of contact heat and infrared radiation, but this could not reproduce the superficiality of the Shroud image, and when his attempt at a replica was examined microscopically, it could be seen that the medullas of the fibrils were colored, unlike the situation with the Shroud, in which only the primary cell wall is discolored.[88]

The evidence suggests that the image does not consist of an added, extraneous substance that has been applied to the surface of the cloth, but instead it consists of a chemical change in the cellulose in some of the outermost surface fibrils.[5,8] The cellulose appears to have been oxidized and dehydrated in a manner akin to an accelerated aging process.[5] There would have to have been an energy source that could provide the energy to cause this change in the cellulose. The transference of this energy would have to have happened either through diffusion or radiation, and these are considered in the next section.

3. Energy Transfer from the Body to the Cloth

The word *energy* is used in this book in the standard way that it is used in science, which relates to the capacity to effect a physical change. Examples of this include heat energy, which has the capacity to raise temperature. Kinetic or "movement" energy has the capacity to produce acceleration or overcome frictional resistance. Gravitational energy also has the potential to do these things. For example, when an object is falling, gravitational potential energy is released, causing the object to accelerate, and some of this energy is used overcoming air resistance so that the object only accelerates until it reaches a "terminal velocity."

The candidates for energy transfer that could have caused the image include:

a) **Chemical energy:** It has been proposed by some that the Shroud image could have been formed by a chemical process known as a Maillard reaction. This was at one time proposed by Los Alamos chemist Raymond Rogers, although apparently he didn't consider that it could account for all the characteristics of the image. Regarding the Shroud image, he suggested that "colour can be produced by reactions between reducing sugars, left on the cloth by the manufacturing procedure, and amines deriving from the decomposition of a corpse."[89] Amines are a type of chemical substance that is produced by a decomposing corpse, although the Shroud image does not show any sign of putrefaction in the body of the man seen on the Shroud. Amines would generally be concentrated around orifices such as the mouth, and yet there is no such enhancement of intensity of the image in these areas on the Shroud. Also, a chemical reaction would depend on chaotic diffusion of decomposition products, and it is difficult to explain how this would produce a focused image, especially as no other examples exist of focused images produced by Maillard reactions in this way. Also, if the image had formed gradually while the cloth had been draped over the recumbent corpse, then we would expect the image to correlate more closely with the bloodstains. When we consider the off-image bloodstain from the elbow and the bloodstains near the hair image that appear to have originated from the sides of the face, we might expect that an image formed by chemical diffusion would conform to a similar alignment of the cloth to the one present when the bloodstains formed and would therefore be a distorted image with the face appearing widened, for example.[14,88]

b) **Heat energy:** Heat can cause chemical changes in linen, but experiments have shown that it is very difficult if not impossible to replicate the superficiality of the changes in this way as heat is rapidly conducted through a linen cloth. Also, an image from a heated bas-relief sculpture will tend to contain the same distortions referred to above. Also, the microscopic and chemical properties of heated linen fibers are different from those of the Shroud's image-bearing fibers.[88,96,97]

c) **Sunlight:** Although rudimentary images can be formed using the camera obscura technique (i.e., an image produced by light shining through a pinhole, as with a pinhole camera), there are numerous issues that dispel the suggestion that the Shroud image could have been formed in this way.[92]

The "medieval photography" hypothesis has been suggested by people such as Professor Nicholas Allen of the Department of Art History and Visual Arts at the University of South Africa, who believes that around four or five hundred years prior to the first documented photograph, a medieval "photographer" managed to produce a light-sensitive emulsion that was then used to coat a piece of linen in a medieval "darkroom."[93] The linen "photographic film" would then have been exposed to light for the first time using a room-sized camera obscura device and a dead body hanging in front of a crystal lens! The first documented technological production of a photographic negative was in 1816 by Joseph Niépce. It was a few years later that he found a way of preserving the images he created. The earliest images he made deteriorated when exposed to light. The earliest remaining photograph of Niépce was made in 1825. Allen argues that the raw materials that could be used to make a photographic emulsion existed in medieval times. However, Barrie Schwortz, the official photographer for STuRP, points out that the raw materials for all modern technological innovations have always been around, but nobody seriously expects to find microwave ovens or nuclear weapons in medieval archaeological excavations.[92] The strongest evidence against the medieval photograph theory is provided by Allen's own experimental results. The Shroud image is unique in that it contains distance-coded information. The intensity of the Shroud image translates proportionately as information relating to the distance between the cloth and the three-dimensional shape of a human body just as it would have done if a body had been contained in the Shroud when the image formed.[52] Although Allen was able to manufacture a photographic negative imprint of a human body on a linen cloth, this image did not have these distance-coding properties.[92] Also, in common with many other postulated mechanisms of image formation, Allen fails to account for the origin of the bloodstains that formed before the image and that provide strong forensic evidence that the Shroud once wrapped the recently deceased body of a man who had been whipped, tortured, and crucified.

d) **Electrical field energy:** Professor Giulio Fanti of the Department of Mechanical Engineering at Padua University has postulated that the Shroud image could be the result of an electrical corona discharge from the body that the Shroud wrapped.[94] A corona discharge occurs when an object has a strong enough electrostatic charge to cause ionization of the surrounding air such that the electrical energy is conveyed away from the object, and under certain circumstances the pattern of electrical discharge can cause an image to be formed of the object. However, electrical discharge has never been known to produce focused images of a body at a distance, and it would seem that there are too many variables relating to the complex patterns of electrical discharge for such an image to be expected to occur with such focus, clarity, and resolution.

e) **Ultraviolet light:** Experiments by physicist Paolo Di Lazzaro and others at ENEA have shown that certain properties of the image, such as the extreme superficiality (around 200 nanometers in thickness) of the image coloration and the lack of fluorescence, can be replicated by short, intense bursts of ultraviolet radiation.[20] I have discussed this in more detail in chapter 2.

Glossary

Chromophore

The part of a chemical structure that is responsible for the color of a substance.

Delayed choice

An experiment in quantum physics whereby the choice of how to make a measurement is delayed until *after* the event that is being measured has occurred.

Determinism

The notion that the future is fixed, regardless of whether it can be predicted.

Entanglement

Quantum entanglement exists when two particles have characteristics or properties that cannot be defined without reference to the other particle even though they are separated in space and time. They may, for example, have opposite spin orientation such that if the orientation of one particle is measured then the other is known (although until either particle is measured there is no fixed outcome for either of them). It is not meaningful to say that the measurement of one must *cause* the outcome of the measurement of the other. Relativistic considerations mean that the particle that appears to be measured *first* may depend upon the observer's frame of reference.

Free will

In this book, free will is *primary causation*. By this I mean that any free act of choice is a fundamental cause in the sense that until the decision is made it is not fixed. The sentient being who makes the decision shapes the present and influences the future through making that decision.

Hidden variables

Hidden-variable theories seek to avoid the indeterminism or randomness of the results of measurements of quantum properties by postulating that there are unknown factors that do determine the outcomes of measurement. These unknown factors would only appear to be random because we do not see these influences.

Materialism

Materialism is the belief that matter and materiality constitute all reality. Many people think that this belief is supported by science. However, as I have demonstrated in this book, science has actually been instrumental in providing copious evidence that materialism is false.

Separability

Quantum separability is effectively the opposite of entanglement. If two particles are separable, then each has properties that do not need to be defined with reference to the other particle. If they are entangled, then they do not have separability until *after* they are measured.

References

This book uses sequentially numbered references. This efficient system allows a single note number to be used for the same resource through-out the entire book. If a lower number appears after a higher one on occasion, this just means that there has been an earlier reference to the same item.

If you struggle to find a given internet source, please try accessing it through my personal website, www.andrewsilverman.co.uk. In addition to my own papers, I provide links to a number of sources used in this book.

1. Anders Sandberg and Nick Bostrom, *Whole Brain Emulation: A Roadmap*, Technical Report #2008-3, Future of Humanity Institute, Oxford University, 2008.
2. Erwin Schrödinger, *What Is Life? with Mind and Matter* (Cambridge: Cambridge University Press, 1967).
3. Andrei Linde, "Universe, Life, Consciousness," available as a pdf file at the Andrei Linde, Professor of Physics, Stanford University website.
4. Andrew Silverman, "The Image on the Shroud: Natural, Manufactured, Miracle, or Something Else?" available on the author's "Light of The Shroud" website or as a pdf from the Shroud of Turin Education and Research Association, Shroud of Turin website.
5. John H. Heller and Alan D. Adler, "A Chemical Investigation of the Shroud of Turin," *Canadian Society of Forensic Science Journal* 14, no. 3 (1981): 81–103.

6. John H. Heller and Alan D. Adler, "Blood on the Shroud of Turin," *Applied Optics* 19, no. 16 (August 15, 1980): 2742–44.

7. Rory Fitzgerald, "A Forger Would Have Needed a Miracle," *The Catholic Herald,* February 17, 2012.

8. Eric J. Jumper et al., "A Comprehensive Examination of the Various Stains and Images on the Shroud of Turin," in *Archaeological Chemistry—III,* ACS Advances in Chemistry Series, edited by J. Lambert (Washington D.C.: American Chemical Society, 1984), 447–76.

9. Eugenia L. Nitowski, "The Field and Laboratory Report of the Environmental Study of the Shroud in Jerusalem 1986," available as a pdf file from the Shroud of Turin Education and Research Association, Shroud of Turin website.

10. David Rolfe, director, *The Silent Witness,* Screenpro Films, 1978.

11. Avinoam Danin, *Botany of the Shroud* (Jerusalem: Danin Publishing, 2010).

12. Ian Wilson, *The Shroud* (New York: Bantam, 2010).

13. Herbert Danby, trans., *The Mishnah,* "Sixth Division, Tohoroth (Cleannesses)" (Oxford: Oxford University Press, 1933), 653–54.

14. Gilbert R. Lavoie, *Unlocking the Secrets of the Shroud* (Allen, Tex.: Thomas More Publishing Company, 1998).

15. Mommsen, Theodor, *Römisches Strafrecht* (Leipzig, Germany: Duncker and Humblot, 1899), 987.

16. E. Mary Smallwood, *The Jews under Roman Rule* (Leiden, the Netherlands: Brill Academic Publishers, 2001), ch. 7.

17. Werner Bulst, "Some Comments on the Turin Shroud after the Carbon Test," *Shroud News,* no. 54 (August 1989).

18. J. M. Rodriguez Almenar, *El Sudario de Oviedo* (Pamplona, Spain: Ediciones Universidad de Navarra, 2000).

19. Mark Guscin, "Recent Historical Investigations on the Sudarium of Oviedo," available as a pdf file from the Shroud of Turin Education and Research Association, Shroud of Turin website, June 1999.

20. P. D. Di Lazzaro, "Superficial and Shroud-like Coloration of Linen by Short Laser Pulses in the Vacuum Ultraviolet," *Applied Optics* 51, no. 36 (December 2012): 8567–78.

21. P. D. Di Lazzaro et al: "Shroud-like Colouration of Linen Fabrics by Far Ultraviolet Radiation: Summary of Results Obtained at the ENEA Frascati Centre 2005–2010," ENEA.

22. John Searle, "Minds, Brains, and Programs," *Behavioural and Brain Sciences* 3, no. 3 (1980): 417–57.

23. Roger Penrose, *The Emperor's New Mind* (Oxford: Oxford University Press, 1989).

24. "Kena Upanishad (a.k.a. Talavakaara Upanishad)," Mumukshu: Easier Way across the Ocean website, July 21, 2009.

25. Kelsey Piper, "The Case for Taking AI Seriously as a Threat to Humanity," Vox, May 8, 2019.

26. Patrick Sawer, "Threat from Artificial Intelligence Not Just Hollywood Fantasy," *Telegraph,* June 27, 2015.

27. Lord Martin Rees, interview by Colin Stuart, "Lord Martin Rees— Astronomer Royal," *BBC Science Focus,* April 13, 2016.

28. Arthur Stanley Eddington, *The Nature of the Physical World* (London: Cambridge University Press, 1928).

29. Eugene Wigner, *Symmetries and Reflections: Scientific Essays* (Cambridge, Mass.: MIT Press, 1967).

30. Andrei Linde, "Why Explore Cosmos and Consciousness?" video of a conversation with Robert Lawrence Kuhn (long version), available at the Closer to Truth website.

31. Stephen Hawking, *A Brief History of Time* (New York: Bantam, 1988).

32. Freeman J. Dyson, "Energy in the Universe," *Scientific American,* September 1971, 51–59.

33. Ian Stevenson, *Children Who Remember Previous Lives* (Charlottesville: University Press of Virginia, 1987).

34. Ciara Curtin, "Fact or Fiction? Living People Outnumber the Dead," *Scientific American* website, March 1, 2007.

35. Andre Linde, *Particle Physics and Inflationary Cosmology* (New York: Harwood, 1990).

36. Bruce Rosenblum and Fred Kuttner, *Quantum Enigma* (Oxford: Oxford University Press, 2011).

37. Alain Aspect, Philippe Grainger, and Gérard Roger, "Experimental Realization of Einstein-Podolsky-Rosen-Bohm *Gedankenexperiment:* A New Violation of Bell's Inequalities," *Physical Review Letters* 49 (1982): 91–94.

38. Simon Gröblacher et al., "An Experimental Test of Non-local Realism," *Nature* 446 (2007): 871–75.

39. John Waller, *Einstein's Luck* (Oxford: Oxford University Press, 2002).

40. Roger Penrose, *Cycles of Time* (London: Bodley Head 2010).

41. "'Antimatter,' Angels & Demons: The Science behind the Story," CERN website, 2011.

42. Renée Weber, *Dialogues with Scientists and Sages: The Search for Unity* (London: Routledge & Kegan Paul, 1986).

43. Lee Smolin, *The Trouble with Physics* (London: Penguin 2007).

44. Max Planck, interview by J. W. N. Sullivan, "Interviews with Great Scientists (Max Planck)," *Observer,* January 25, 1931, 17.

45. Sam Parnia, "Do Reports of Consciousness during Cardiac Arrest Hold the Key to Discovering the Nature of Consciousness?" *Medical Hypotheses* 69, no. 4 (2007): 933–37.

46. Pim Van Lommel et al., "Near Death Experiences in Survivors of Cardiac Arrest: A Prospective Study in the Netherlands," *The Lancet* 358, no. 9298 (December 15, 2001): P2039–45.

47. Sam Parnia et al., "AWARE—AWAreness during REsuscitation—A Prospective Study," *Resuscitation* 85, no. 12 (December 2014): 1799–1805.

48. Mahendra Perera et al., *Making Sense of Near-Death Experiences: A Handbook for Clinicians* (London: Jessica Kingsley Publishers, 2012).

49. Roger Penrose, *Shadows of the Mind* (Oxford: Oxford University Press, 1994).

50. Gilbert R. Lavoie, "Turin Shroud: A Medical Forensic Study of Its Blood Marks and Image," in *Proceedings of the International Workshop on the Scientific Approach to the Acheiropoietos Images,* ENEA, Frascati, Italy, May 4–6, 2010, available as a pdf file on the International Workshop on the Scientific Approach to the Acheiropoietos Images website.

51. John Jackson, "Is the Image on the Shroud Due to a Process Heretofore Unknown to Modern Science?" *Shroud Spectrum International* 34 (1990): 3–29.

52. John P. Jackson, Eric J. Jumper, and William R. Ercoline, "Three-Dimensional Characteristics of the Shroud Image," in *IEEE 1982 Proceedings of the International Conference on Cybernetics and Society,* Seattle, Wash., October 28–30, 1982, 559–75.

53. Johnjoe McFadden and Jim Al-Khalili, *Life on the Edge: The Coming of Age of Quantum Biology* (New York: Bantam, 2014).

54. Jack Tuszinski, *The Emerging Physics of Consciousness* (Berlin: Springer-Verlag, 2006).

55. Jonas Mureika and Dejan Stojkovic, "Detecting Vanishing Dimensions via Primordial Gravitational Wave Astronomy," *Physical Review Letters* 106, no. 101101 (March 8, 2011).

56. Werner Heisenberg, *Physics and Philosophy: The Revolution in Modern Science* (New York: Harper and Row, 1962).

57. John Archibald Wheeler, "Hermann Weyl and the Unity of Knowledge," *American Scientist* 74 (July–August 1986): 366–75, also available as a pdf file on the Weylmann website.

58. John Archibald Wheeler, "Information, Physics, Quantum: The Search for Links," in *Complexity, Entropy, and the Physics of Information,* edited by W. Zurek (Boston: Addison-Wesley, 1990).

59. Walter Isaacson, *Einstein: His Life and Universe* (New York: Simon and Schuster, 2007).

60. Lawrence Krauss, *A Universe from Nothing: Why There Is Something Rather than Nothing* (New York: Free Press, 2012).

61. James Jeans, *The Mysterious Universe* (London: MacMillan, 1932).

62. Rupert Sheldrake, *The Science Delusion* (Philadelphia, Pa.: Coronet, 2012).

63. "Butterfly Effect," Wikipedia.

64. Roger Nelson, "Detecting Mass Consciousness: Effects of Globally Shared Attention and Emotion," *Journal of Cosmology* 14 (2011): 4616–32.

65. Roger Colbeck et al., "No Extension of Quantum Theory Can Have Improved Predictive Power," *Nature Communications* 2, no. 411 (2011).

66. Vincent Jacques et al., "Experimental Realization of Wheeler's Delayed-Choice Gedanken Experiment," *Science* 315, no. 5814 (February 2007): 966–68.

67. Alice Calaprice, ed., *The Expanded Quotable Einstein* (Princeton, N.J.: Princeton University Press, 2000).

68. Lee Smolin, *Time Reborn: From the Crisis of Physics to the Future of the Universe* (London: Penguin, 2013).

69. Erwin Schrödinger, "Die gegenwärtige Situation in der Quantenmechanik [The present situation in quantum mechanics]," *Naturwissenschaften* 23 no. 48 (1935): 807–12.

70. Erwin Schrödinger, interview by J. W. N. Sullivan, "Interviews with Great Scientists (Erwin Schrödinger)," *Observer,* January 11, 1931, 16.

71. Palle Yourgrau, *A World without Time: The Forgotten Legacy of Gödel and Einstein* (London: Penguin, 2007).

72. Stephen Hawking, "The Chronology Protection Conjecture," *Physical Review D* 46 (July 1992): 603–11.

73. John Archibald Wheeler, *Quantum Theory and Measurement* (Princeton, N.J.: Princeton University Press, 1983).

74. Erwin Schrödinger, *My View of the World* (Woodbridge, Conn.: Ox Bow Press, 1983).

75. Paul Schillp, ed., *The Philosophy of Rudolf Carnap: Intellectual Autobiography* (Chicago: Open Court, 1963).

76. Sean Carroll, *From Eternity to Here* (London: Penguin, 2010).

77. University of Cambridge Centre for the Study of Existential Risk website.

78. University of Cambridge, "Professor Lord Martin Rees," Cambridge Forum for Sustainability and the Environment website.

79. P. E. Damon et al., "Radiocarbon Dating of the Shroud of Turin" *Nature* 337, no. 6208 (1989): 611–15.

80. Giulio Fanti et al., "A Robust Statistical Analysis of the 1988 Turin Shroud Radiocarbon Dating Results" in *Proceedings of the International Workshop on the Scientific Approach to the Acheiropoietos Images,* ENEA, Frascati, Italy, May 4–6, 2010.

81. Raymond Rogers, "Studies on the Radiocarbon Sample from the Shroud of Turin," *Thermochimica Acta* 425, no. 1 (January 2005): 189–94.

82. M. Sue Benford and Joseph G. Marino "Discrepancies in the Radiocarbon Dating Area of the Turin Shroud," *Chemistry Today* 26, no. 4 (July–August 2008): 4–12.

83. O. Petrosillo and E. Marinelli, *The Enigma of the Shroud* (San Gwann, Malta: PEG Publications, 1996).

84. Joseph G. Marino, *Wrapped Up in the Shroud* (St. Louis, Mo.: Cradle Press, 2011).

85. David Rolfe, producer, *Shroud of Turin: Material Evidence,* BBC Television, 2008.

86. Thibault Heimburger and Giulio Fanti, "Scientific Comparison between the Turin Shroud and the First Handmade Whole Copy," *Proceedings of the International Workshop on the Scientific Approach to the Acheiropoietos Images,* ENEA, Frascati, Italy, May 4–6, 2010, available as a pdf file on the International Workshop on the Scientific Approach to the Acheiropoietos Images website.

87. Mark Guscin, "The Making of a New Shroud Documentary," *Shroud Newsletter,* no. 74 (December 2011), from the British Society for the Turin Shroud, available as a pdf file from the Shroud of Turin Education and Research Association, Shroud of Turin website.

88. Paolo Di Lazzaro, "Could a Burst of Radiation Create a Shroud-like Coloration? Summary of 5-Years Experiments at ENEA Frascati," available as a pdf from the Shroud of Turin Education and Research Association, Shroud of Turin website.

89. Raymond Rogers and Anna Arnoldi, "The Shroud of Turin: An Amino-Carbonyl Reaction (Maillard Reaction) May Explain the Image Formation," *Melanoidins* 4 (2003): 106–13.

90. L. A. Schwalbe and R. N. Rogers. "Physics and Chemistry of the Shroud of Turin: A Summary of the 1978 Investigation," *Analytica Chimica Acta* 135, no. 1 (February 1982): 3–49.

91. V. D. Miller and S. F. Pellicori. "Ultraviolet Fluorescence Photography of the Shroud of Turin," *Journal of Biological Photography* 49, no. 3 (July 1981): 71–85.

92. Barrie M. Schwortz, "Is the Shroud of Turin a Medieval Photograph? A Critical Examination of the Theory," available as a pdf file from the Shroud of Turin Education and Research Association, Shroud of Turin website.

93. N. P. L Allen, "Is the *Shroud of Turin* the First Recorded Photograph?" *South African Journal of Art History* 11 (1993): 23–32.

94. Giulio Fanti, "Body Image Formation Hypotheses Based on Corona Discharge: Discussion," paper presented at the International Conference on the Shroud of Turin, "Perspectives of a Multifaceted Enigma," Columbus, Ohio, August 14–17, 2008, available as a pdf file on the Shroud Science Group International Conference website.

Bibliography

A bibliography has been provided to assist readers looking to locate sources used in this book. Sequenced alphabetically by author surname, it should alleviate the need to search through the references list.

Allen, N. P. L. "Is the Shroud of Turin the First Recorded Photograph?" *South African Journal of Art History* 11 (1993): 23–32.

"'Antimatter.' Angels & Demons: The Science behind the Story." CERN website, 2011.

Aspect, Alain, Philippe Grainger, and Gérard Roger. "Experimental Realization of Einstein-Podolsky-Rosen-Bohm *Gedankenexperiment*: A New Violation of Bell's Inequalities." *Physical Review Letters* 49 (1982): 91–94.

Benford, M. Sue, and Joseph G. Marino. "Discrepancies in the Radiocarbon Dating Area of the Turin Shroud." *Chemistry Today* 26, no. 4 (July–August 2008): 4–12.

Bulst, Werner. "Some Comments on the Turin Shroud after the Carbon Test." *Shroud News,* no. 54 (August 1989).

Calaprice, Alice, ed. *The Expanded Quotable Einstein.* Princeton, N.J.: Princeton University Press, 2000.

Carroll, Sean. *From Eternity to Here.* London: Penguin, 2010.

Colbeck, Roger and Renator Renner. "No Extension of Quantum Theory Can Have Improved Predictive Power. *Nature Communications* 2, no. 411 (2011).

Curtin, Ciara. "Fact or Fiction? Living People Outnumber the Dead." *Scientific American* website, March 1, 2007.

Damon, P. E., D. J. Donahue, B. H. Gore, A. L. Hatheway, A. J. T. Jull, T. W. Linick, P. J. Sercel, et al. "Radiocarbon Dating of the Shroud of Turin." *Nature* 337, no. 6208 (1989): 611–15.

Danby, Herbert, trans. *The Mishnah.* "Sixth Division, Tohoroth ('Cleannesses')," 653–54. Oxford: Oxford University Press, 1933.

Danin, Avinoam. *Botany of the Shroud.* Jerusalem: Danin Publishing, 2010.

Di Lazzaro, Paolo D. "Could a Burst of Radiation Create a Shroud-like Coloration? Summary of 5-Years Experiments at ENEA Frascati." Available as a pdf file from the Shroud of Turin Education and Research Association, Shroud of Turin website.

———. "Superficial and Shroud-like Coloration of Linen by Short Laser Pulses in the Vacuum Ultraviolet." *Applied Optics* 51, no. 36 (December 2012): 8567–78.

Di Lazzaro, Paolo D., Danielle Murra, Enrico Nichelatti, Antonio Santoni, Giuseppe Baldacchini. "Shroud-like Colouration of Linen Fabrics by Far Ultraviolet Radiation: Summary of Results Obtained at the ENEA Frascati Centre 2005–2010," ENEA.

Dyson, Freeman J. "Energy in the Universe." *Scientific American,* September 1971, 51–59.

Eddington, Arthur Stanley. *The Nature of the Physical World.* London: Cambridge University Press, 1928.

Fanti, Giulio. "Body Image Formation Hypotheses Based on Corona Discharge: Discussion." Paper presented at the International Conference on the Shroud of Turin, "Perspectives of a Multifaceted Enigma," Columbus, Ohio, August 14–17, 2008. Available as a pdf file on the Shroud Science Group International Conference website.

Fitzgerald, Rory. "A Forger Would Have Needed a Miracle." *The Catholic Herald,* February 17, 2012.

Gröblacher, Simon, Tomasz Paterek, Rainer Kaltenbaek, Časlav Brukner, Marek Żukowski, Markus Aspelmeyer, and Anton Zeilinger. "An Experimental Test of Non-local Realism." *Nature* 446 (2007): 871–75.

Guscin, Mark. "The Making of a New Shroud Documentary." *Shroud Newsletter,* no. 74 (December 2011), from the British Society for the Turin Shroud. Available as a pdf file from the Shroud of Turin Education and Research Association, on the Shroud of Turin website.

———. "Recent Historical Investigations on the Sudarium of Oviedo." Available as a pdf file from the Shroud of Turin Education and Research Association, Shroud of Turin website. 1999.

Hawking, Stephen. *A Brief History of Time*. New York: Bantam, 1988.

———. "The Chronology Protection Conjecture." *Physical Review D* 46 (July 1992): 603–11.

Heimburger, Thibault, and Giulio Fanti. "Scientific Comparison between the Turin Shroud and the First Handmade Whole Copy." *Proceedings of the International Workshop on the Scientific Approach to the Acheiropoietos Images,* ENEA, Frascati, Italy, May 4–6, 2010. Available as a pdf file on the International Workshop on the Scientific Approach to the Acheiropoietos Images website.

Heisenberg, Werner. *Physics and Philosophy: The Revolution in Modern Science*. New York: Harper and Row, 1962.

Heller, John H., and Alan D. Adler. "Blood on the Shroud of Turin." *Applied Optics* 19, no. 16 (August 15, 1980): 2742–44.

———. "A Chemical Investigation of the Shroud of Turin." *Canadian Society of Forensic Science Journal* 14, no. 3 (1981): 81–103.

Isaacson, Walter. *Einstein: His Life and Universe*. New York: Simon and Schuster, 2007.

Jackson, John. "Is the Image on the Shroud Due to a Process Heretofore Unknown to Modern Science?" *Shroud Spectrum International* 34 (1990): 3–29.

Jackson, John P., Eric J. Jumper, and William R. Ercoline. "Three-Dimensional Characteristics of the Shroud Image." In *IEEE 1982 Proceedings of the International Conference on Cybernetics and Society,* Seattle, Wash., October 28–30, 1982, 559–75.

Jacques, Vincent, E. Wu, Frédéric Grosshans, François Treussart, Philippe Grangier, Alain Aspect, and Jean-François Roch. "Experimental Realization of Wheeler's Delayed-Choice Gedanken Experiment." *Science* 315, no. 5814 (February 2007): 966–68.

Jeans, James. *The Mysterious Universe*. London: MacMillan, 1932.

Jumper, Eric J., Alan D. Adler, John P. Jackson, Samuel F. Pellicori, John H. Heller, and James R. Druzik. "A Comprehensive Examination of the Various Stains and Images on the Shroud of Turin." In *Archaeological Chemistry—III*. Advances in Chemistry Series, edited by J. Lambert, 447–76. Washington D.C.: American Chemical Society, 1984.

"Kena Upanishad (a.k.a. Talavakaara Upanishad)." Mumukshu: Easier Way across the Ocean website. July 21, 2009.

Krauss, Lawrence. *A Universe from Nothing: Why There Is Something Rather than Nothing.* New York: Free Press, 2012.

Lavoie, Gilbert R. "Turin Shroud: A Medical Forensic Study of Its Blood Marks and Image." In *Proceedings of the International Workshop on the Scientific Approach to the Acheiropoietos Images,* ENEA, Frascati, Italy, May 4–6, 2010. Available as a pdf file on the International Workshop on the Scientific Approach to the Acheiropoietos Images website.

———. *Unlocking the Secrets of the Shroud.* Allen, Tex.: Thomas More Publishing Company, 1998.

Linde, Andrei. *Particle Physics and Inflationary Cosmology.* New York: Harwood, 1990.

———. "Universe, Life, Consciousness." Available as a pdf file at the Andrei Linde, Professor of Physics, Stanford University website.

———. "Why Explore Cosmos and Consciousness?" Video of a conversation with Robert Lawrence Kuhn. Available at the Closer to Truth website.

London School of Economics website.

Marino, Joseph G. *Wrapped Up in the Shroud.* St. Louis, Mo: Cradle Press, 2011.

McFadden, Johnjoe, and Jim Al-Khalili. *Life on the Edge: The Coming of Age of Quantum Biology.* New York: Bantam, 2014.

Miller, V. D., and S. F. Pellicori. "Ultraviolet Fluorescence Photography of the Shroud of Turin." *Journal of Biological Photography* 49, no. 3 (July 1981): 71–85.

Mommsen, Theodor. *Römisches Strafrecht.* Leipzig, Germany: Duncker and Humblot, 1899.

Mureika, Jonas and Dejan Stojkovic. "Detecting Vanishing Dimensions via Primordial Gravitational Wave Astronomy." *Physical Review Letters* 106, no. 101101 (March 8, 2011).Nelson, Roger. "Detecting Mass Consciousness: Effects of Globally Shared Attention and Emotion." *Journal of Cosmology* 14 (2011): 4616–32.

Nitowski, Eugenia L. "The Field and Laboratory Report of the Environmental Study of the Shroud in Jerusalem 1986." Available as a pdf file from the Shroud of Turin Education and Research Association, on the Shroud of Turin website.

Parnia, Sam. "Do Reports of Consciousness during Cardiac Arrest Hold the

Key to Discovering the Nature of Consciousness?" *Medical Hypotheses* 69, no. 4 (2007): 933–37.

Parnia, Sam, Ken Spearpoint, Gabriele de Vos, Peter Fenwick, Diana Goldberg, Jie Yang, Jiawen Zhu, et al. "AWARE—AWAreness during REsuscitation—A Prospective Study." *Resuscitation* 85, no. 12 (December 2014): 1799–1805.

Penrose, Roger. *Cycles of Time.* London: Bodley Head, 2010.

———. *The Emperor's New Mind.* Oxford: Oxford University Press, 1989.

———. *Shadows of the Mind.* Oxford: Oxford University Press, 1994.

Perera, Mahendra, Karuppiah Jagadheesan, and Anthony Peake, eds. *Making Sense of Near-Death Experiences: A Handbook for Clinicians.* London: Jessica Kingsley Publishers, 2012.

Petrosillo, O., and E. Marinelli. *The Enigma of the Shroud.* San Gwann, Malta: PEG Publications, 1996.

Piper, Kelsey. "The Case for Taking AI Seriously as a Threat to Humanity." Vox, May 8, 2019.

Planck, Max, interview by J. W. N. Sullivan. "Interviews with Great Scientists (Max Planck)." *Observer,* January 25, 1931, 17.

Rees, Lord Martin, interview by Colin Stuart. "Lord Martin Rees—Astronomer Royal." *BBC Science Focus,* April 13, 2016. Available on the *BBC Science Focus* website.

Rodriguez Almenar, J. M. *El Sudario de Oviedo.* Pamplona, Spain: Ediciones Universidad de Navarra, 2000.

Rogers, Raymond. "Studies on the Radiocarbon Sample from the Shroud of Turin." *Thermochimica Acta* 425, no. 1 (January 2005): 189–94.

Rogers, Raymond, and Anna Arnoldi. "The Shroud of Turin: An Amino-Carbonyl Reaction (Maillard Reaction) May Explain the Image Formation." *Melanoidins* 4 (2003): 106–13.

Rolfe, David, producer. *Shroud of Turin: Material Evidence.* BBC Television, 2008.

———, director. *The Silent Witness,* 1978.

Rosenblum, Bruce, and Fred Kuttner. *Quantum Enigma.* Oxford: Oxford University Press, 2011.

Sandberg, Anders and Nick Bostrom. *Whole Brain Emulation: A Roadmap,* Technical Report #2008-3, Future of Humanity Institute, Oxford University, 2008.

Sawer, Patrick. "Threat from Artificial Intelligence Not Just Hollywood Fantasy." *Telegraph,* June 27, 2015.

Schillp, Paul, ed. *The Philosophy of Rudolf Carnap: Intellectual Autobiography.* Chicago: Open Court, 1963.

Schrödinger, Erwin. "Die gegenwärtige Situation in der Quantenmechanik" [The present situation in quantum mechanics]. *Naturwissenschaften* 23, no. 48 (1935): 807–12.

———. *My View of the World.* Woodbridge, Conn.: Ox Bow Press, 1983.

———. *What Is Life?* London: Cambridge University Press, 1967.

Schrödinger, Erwin, interview by J. W. N. Sullivan. *Observer,* January 11, 1931, 16.

Schwalbe, L. A., and R. N. Rogers. "Physics and Chemistry of the Shroud of Turin: A Summary of the 1978 Investigation." *Analytica Chimica Acta* 135, no. 1 (February 1982): 3–49.

Schwortz, B. M. "Is the Shroud of Turin a Medieval Photograph? A Critical Examination of the Theory." Available as a pdf file from the Shroud of Turin Education and Research Association, on the Shroud of Turin website.

Science Made Simple, Inc. "What is Static Electricity?" Science Made Simple website.

Searle, John. "Minds, Brains, and Programs." *Behavioural and Brain Sciences* 3, no. 3 (1980): 417–57.

Sheldrake, Rupert. *The Science Delusion.* Philadelphia, Pa.: Coronet, 2012.

Silverman, Andrew. "The Image on the Shroud: Natural, Manufactured, Miracle, or Something Else?" Available as a pdf file from the Shroud of Turin Education and Research Association, Shroud of Turin website.

Smallwood, E. Mary. *The Jews under Roman Rule.* Leiden, the Netherlands: Brill Academic Publishers, 2001.

Smolin, Lee. *Time Reborn: From the Crisis of Physics to the Future of the Universe.* London: Penguin, 2013.

———. *The Trouble with Physics.* London: Penguin, 2007.

Stevenson, Ian. *Children Who Remember Previous Lives.* Charlottesville: University Press of Virginia, 1987.

Tuszinski, Jack. *The Emerging Physics of Consciousness.* Berlin: Springer-Verlag, 2006.

University of Cambridge Centre for the Study of Existential Risk website.

University of Cambridge. "Professor Lord Martin Rees." Cambridge Forum for Sustainability and the Environment website.

Van Lommel, Pim, Ruud van Wees, Vincent Meyers, and Ingrid Elfferich. "Near-Death Experiences in Survivors of Cardiac Arrest: A Prospective Study in the Netherlands." *Lancet* 358, no. 9298 (December 15, 2001) P2039–45.

Waller, John. *Einstein's Luck*. Oxford: Oxford University Press, 2002.

Weber, Renée. *Dialogue with Scientists and Sages: The Search for Unity*. London: Routledge & Kegan Paul, 1986.

Wheeler, John Archibald. "Hermann Weyl and the Unity of Knowledge." *American Scientist* 74 (July–August 1986): 366–75. Available as a pdf file at the Weylmann website.

———. "Information, Physics, Quantum: The Search for Links." In *Complexity, Entropy, and the Physics of Information,* edited by W. Zurek. Boston: Addison-Wesley, 1990.

———. *Quantum Theory and Measurement*. Princeton, N.J.: Princeton University Press, 1983.

Wigner, Eugene. *Symmetries and Reflections: Scientific Essays*. Cambridge, Mass.: MIT Press, 1967.

Wilson, Ian. *The Shroud*. New York: Bantam, 2010.

Yourgrau, Palle. *A World without Time: The Forgotten Legacy of Gödel and Einstein*. London: Penguin, 2007.

Index

Page numbers in *italic* type refer to illustrations.